Welding Deformation and Residual Stress Prevention

Welding Deformation and Residual Stress Prevention

Yukio Ueda
Joining and Welding Research Institute
Osaka University, Japan

Hidekazu Murakawa
Joining and Welding Research Institute
Osaka University, Japan

Ninshu Ma
Joining and Welding Research Institute
Osaka University, Japan

ELSEVIER

AMSTERDAM • BOSTON • HEIDELBERG • LONDON
NEW YORK • OXFORD • PARIS • SAN DIEGO
SAN FRANCISCO • SINGAPORE • SYDNEY • TOKYO

Butterworth-Heinemann is an imprint of Elsevier

Butterworth-Heinemann is an imprint of Elsevier
225 Wyman Street, Waltham, MA 02451, USA
The Boulevard, Langford Lane, Kidlington, Oxford, OX5 1GB, UK

Notices

Knowledge and best practice in this field are constantly changing. As new research and experience broaden our
understanding, changes in research methods, professional practices, or medical treatment may become necessary.

Practitioners and researchers must always rely on their own experience and knowledge in evaluating and using
any information, methods, compounds, or experiments described herein. In using such information or methods
they should be mindful of their own safety and the safety of others, including parties for whom they have a
professional responsibility.

To the fullest extent of the law, neither the Publisher nor the authors, contributors, or editors assume any liability
for any injury and/or damage to persons or property as a matter of products liability, negligence or otherwise, or
from any use or operation of any methods, products, instructions, or ideas contained in the material herein.

Library of Congress Cataloging-in-Publication Data
Ueda, Yukio.
  Welding deformation and residual stress prevention / Yukio Ueda, Hidekazu Murakawa, Ninshu Ma. – 1st ed.
    p. cm.
  ISBN 978-0-12-394804-5
1. Residual stresses. 2. Welded joints–Defects. 3. Deformations (Mechanics) I. Murakawa, Hidekazu.
II. Ma, Ninshu. III. Title.
  TA492.W4U33 2012
  671.5'20422–dc23
                                    2011052964

British Library Cataloguing-in-Publication Data
A catalogue record for this book is available from the British Library.

For information on all Butterworth-Heinemann publications
visit our website at *www.elsevierdirect.com*

Typeset by: diacriTech, Chennai, India

Printed in the United States of America
12 13 14 15   10 9 8 7 6 5 4 3 2 1

Contents

Preface   xi
List of Symbols   xv

1 Introduction to Welding Mechanics

1.1 Basic Concepts of Welding and Welding Mechanics   1
1.2 Process of the Production of Residual Stress and Inherent Strain   8
    1.2.1 Illustrative Model   8
    1.2.2 Heating of Free Bar C   12
    1.2.3 Heating of Bar C Fixed at Both Ends   13
    1.2.4 Heating Bar C When It Is Connected to a Movable
         Rigid Body   21
1.3 Reproduction of Residual Stress by Inherent Strain
    and Inverse Analysis for Inherent Strain   27
    1.3.1 Reproduction of Residual Stress by Inherent Strain   27
    1.3.2 Inverse Analysis for Inherent Strain   28
1.4 Numerical Examples of Residual Stress, Inherent Strain, and
    Inherent Displacement   29
    1.4.1 Low-Temperature Heating Process   30
    1.4.2 Medium-Temperature Heating   31
    1.4.3 High-Temperature Heating   33
References   34

2 Introduction to Measurement and Prediction of Residual Stresses with the Help of Inherent Strains

2.1 Inherent Strains and Resulting Stresses   36
    2.1.1 Displacement–Strain Relation (Compatibility)   38
    2.1.2 Stress–Strain Relation (Constitution Equation) 39
    2.1.3 Equilibrium Condition (Equilibrium Equation)   39
2.2 Measured Strains in Experiments and Inherent Strains   40
2.3 Effective and Noneffective Inherent Strains   41
2.4 Determination of Effective Inherent Strains from
    Measured Residual Stresses   42
2.5 Most Probable Value of Effective Inherent Strain
    and Accuracy of the Measurement of Residual Stress   45

2.6   Derivation of Elastic Response Matrix   46
2.7   Measuring Methods and Procedures of Residual Stresses
      in Two- and Three-Dimensional Models   48
      2.7.1   Measurement of Two-Dimensional Residual Stresses
             Induced in the Butt-Welded Joint of a Plate   48
      2.7.2   Measurement of Three-Dimensional Residual Stresses
             Induced in Thick Plates   49
2.8   Prediction of Welding Residual Stresses   52
References   52

3   Mechanical Simulation of Welding

3.1   Heat Flow and Temperature During Welding   55
      3.1.1   Heat Supply, Diffusion, and Dissipation   56
      3.1.2   Simple Heat Flow Model   59
      3.1.3   Differences in Material Properties   63
      3.1.4   Change of Material Properties with Temperature   64
      3.1.5   Characteristic Temperature and Length   69
      3.1.6   Simple Method for Solving the Heat Conduction Problem   74
      3.1.7   Summary   79
3.2   Basic Concepts of Mechanical Problems in Welding   79
      3.2.1   Classification of Problems According to Dimensions   80
      3.2.2   Variables and Equations Used to Describe Mechanical
             Problems   81
      3.2.3   Deformation and Stress in the Three-Bar Model   84
      3.2.4   Stress-Strain Relation in Welding   87
      3.2.5   Thermal Visco-Elasto-Plastic Problem in Three-Bar Model   94
      3.2.6   Closing Remarks   96
References   97

4 The Finite Element Method

4.1   Finite Element Method as a Powerful Tool for a Variety
      of Problems   99
4.2   Types of Problems and the Corresponding Basic Equations   101
4.3   Basic Concepts of the Variational Principle   103
4.4   How to Solve a Problem with More than One Element   104
      4.4.1   Equilibrium Equation of a Bar   104
      4.4.2   Equilibrium Equations of Two Bars   105
4.5   Incremental Method for Nonlinear Problems   108
4.6   Simple Examples of Analyzing Thermal
      Elastic-Plastic-Creep Behavior   109
      4.6.1   Bar Fixed at Both Ends Under a Thermal Cycle   109
      4.6.2   Thermal Elastic Behavior of a Bar Fixed at Both Ends   112
      4.6.3   Thermal Elastic-Plastic Behavior of a Bar Fixed
             at Both Ends   113
      4.6.4   Thermal Elastic-Plastic Creep Behavior of a Bar Fixed
             at Both Ends   114

4.7 Basic Theoretical Solutions to Validate Results Obtained
by the FEM   115
    4.7.1 Temperature Distribution Due to a Concentrated
          Heat Source   115
    4.7.2 Temperature Distribution on a Butt-Welded Joint of a
          Thin Plate   116
    4.7.3 Temperature Distribution on a Butt-Welded Joint of a
          Thick Plate   119
    4.7.4 Inherent Strain Distribution on a Butt-Welded Joint of
          a Thin Plate   120
4.8 Flow of Analysis for Welding Deformation and Residual Stress 122
4.9 Checklist for Rational Simulation   124
    4.9.1 Checklist for Preparation of Input Data   124
    4.9.2 Checklists for the Results of Simulation   125
4.10 Troubleshooting for Problems Experienced in Computation   126
    4.10.1 Troubleshooting for Common Problems in Heat Conduction
           Analysis and Stress Analysis   126
    4.10.2 Troubleshooting for Heat Conduction Analysis   126
    4.10.3 Troubleshooting for Thermal Elastic-Plastic Analysis 128
References   129

5 Q&A for FEM Programs

5.1 Q&A for Program Introduction   131
5.2 Q&A for Welding Heat Conduction Program heat2d.exe   135
5.3 Q&A for Thermal Elastic-Plastic Creep Program tepc2d.exe   140
5.4 Q&A for the Inherent Strain–Based Program inhs2d.exe   142
5.5 Q&A for Postprocessing Program awsd.exe   144
5.6 Q&A for Sample Data   152
Reference   167

6 Simulation Procedures for Welding Heat Conduction,
  Welding Deformation, and Residual Stresses Using the
  FEM Programs Provided on the Companion Website

6.1 Simulation Steps Using the Welding Heat Conduction
    FEM Program   170
    6.1.1 Purpose and Simulation Conditions   170
    6.1.2 Preparation of Input File   171
    6.1.3 Steps to Execute heat2d.exe for Welding Heat Conduction 176
    6.1.4 Viewing Results Using Postprocessing Program   176
6.2 Simulation Steps Using the Thermal Elastic-Plastic Creep
    FEM Program   178
    6.2.1 Purpose and Simulation Conditions   178
    6.2.2 Preparation of Input File   179
    6.2.3 Prepare the Node Temperature File   185
    6.2.4 Steps to Execute the FEM Program   185
    6.2.5 Viewing Results Using the Postprocessing Program   185

6.3   Simulation Steps Using the Inherent Strain FEM Program   188
       6.3.1   Purpose and Simulation Conditions   188
       6.3.2   Preparation of Input File   189
       6.3.3   Steps to Execute the Program   192
       6.3.4   Comparison of the Results of the Inherent Strain Method
               and the Thermal Elastic-Plastic Method   192
6.4   Numerical Experiment for Residual Stress Measurement
      Using the Inherent Strain FEM Program   193
       6.4.1   Purpose and Simulation Conditions   193
       6.4.2   Preparation of Input File   194
       6.4.3   Steps to Execute the FEM Program for Residual
               Stress and Deformation   199
       6.4.4   Comparison of the Results of the Inherent Strain
               Method and the Thermal Elastic-Plastic Method   199
6.5   Computation Steps for the Prediction of Residual Stresses
      by the Inherent Strain Method   200
       6.5.1   Purpose and Simulation Conditions   200
       6.5.2   Preparation of Input File and Prediction Formula of
               Inherent Strain   200
       6.5.3   Steps to Execute inhs2d.exe for Residual Stress Computation   206
       6.5.4   Comparison of the Results by the Inherent Strain Method
               and the Thermal Elastic-Plastic Method   206
Reference   207

7   Strategic Simulation Analyses for Manufacturing
    Problems Related to Welding

7.1   Cold Cracking at the First Pass of a Butt-Welded Joint Under
      Mechanical Restraint   210
7.2   Cold Cracking of Slit Weld   213
7.3   Analysis of Welding Residual Stress of Fillet Welds
      for Prevention of Fatigue Cracks   216
       7.3.1   Residual Stresses by Three-Dimensional Analysis   216
       7.3.2   Comparison of Residual Stresses by Two-Dimensional
               and Three-Dimensional Analyses   216
       7.3.3   Comparison of Residual Stresses in Single-Pass
               and Multipass Welds   219
7.4   Multipass-Welded Corner Joints and Weld Cracking   221
       7.4.1   Experiment and Result   221
       7.4.2   Residual Stresses by Thermal Elastic-Plastic Analysis   221
       7.4.3   Effects of Welding Residual Stress and Geometry
               of Edge Preparation on Initiation of Welding Cracks   223
7.5   Analysis of Transient and Residual Stresses of Multipass
      Welding of Thick Plates in Relation to Cold Cracks,
      Under-Bead Cracks, Etc.   227
       7.5.1   Specimens and Conditions for Theoretical Analysis   227
       7.5.2   Characteristics of Welding Residual Stress Distributions
               and Production Process   230

7.6 Improvement of Residual Stresses of a Circumferential
      Joint of a Pipe by Heat-Sink Welding   231
7.7 Prediction of Deformation Produced by Line Heating   235
7.8 Simulation of Resistance Spot Welding Process   236
7.9 Prediction of Welding Distortion Produced in Large
      Plate Structures   239
References   244

**Appendix A Residual Stress Distributions in Typical
              Welded Joints   247**

A.1 Residual Stresses in Base Metals   248
      A.1.1 Residual Stress in Thermo-Mechanical Control Process
             (TMCP) Steel   248
      A.1.2 Residual Stress in TMCP Steel Induced by Bead Weld   250
      A.1.3 Explosive Clad Steel   251
      A.1.4 Cylindrical Thick Plate by Cold Bending   252
A.2 Residual Stresses in Welded Joints of Plates;
      in 2-Dimensional   253
      A.2.1 Butt-Welded Joints; Classification of Patterns of Residual
             Stress Distributions   253
      A.2.2 Long Butt-Welded Joint, Prediction Equation   256
      A.2.3 Built-Up Members of T Shape and I Shape   257
      A.2.4 Built-Up Member of T Shape, Experiment   259
      A.2.5 Residual Stress and Inherent Displacement Induced
             by Slit Welds   260
A.3 Mulltipass Butt Welds of Thick Plates; 3-Dimensional   263
      A.3.1 Multipass Butt Welds of Thick Plates, Classification 263
      A.3.2 Multipass Butt Welds of Thick Plate, Experiment   264
A.4 Electron Beam Welding, Thick Plate   265
A.5 First Bead of Butt Joint; RCC (Rigidly Restrained Cracking)
      Test Specimen   267
A.6 Multipass-Welded Corner Joint   268
A.7 Fillet Welds: 3-Dimensional   269
      A.7.1 Single Fillet Welds   269
      A.7.2 Fillet Welds at the Joint of Web and Flange   270
A.8 Repair Weld of Thick Plate   271
A.9 Circumferential Welded Joint of Pipes   274
      A.9.1 Circumferential Welded Joint of Pipes—Heat-Sink
             Welding   274
      A.9.2 Penetrating Pipe Joints in Nuclear Reactor   279
References   281

**Appendix B Published material properties for thermal
              elastic-plastic FEM analysis**
              Located on the companion website at
              booksite.elsevier.com/9780123948045

Appendix C Theory of three-dimensional thermal
elastic-plastic creep analysis
Located on the companion website at
booksite.elsevier.com/9780123948045

Contents of Programs and Data on the Companion
Website   283

Index   285
About the Authors   291

Since arc welding was invented in the late 19th century, welding has been widely used as an essential technology for metal construction. Generally, welding produces welding deformation and residual stress in products, which influences their quality and performance. Although many engineers and researchers have made great efforts to control these incidents, they have remained engineering problems.

Although the fundamental characteristics of arc welding have been clarified and basic theories have been established, the phenomenon is complicated and interdisciplinary and it has been difficult to predict incidents accurately.

In fact, welding begins the instant of heat input of high density by arc welding. As this heat induces melting of the metal and conducts in the joints, local expansion and shrinkage in the joints result in welding deformation and residual stresses.

After the personal computer was invented, new computational theories and methods were developed, one of which is the finite element method (FEM). This method is a powerful numerical analysis tool used to solve complex problems. The authors' research group has developed many computational methods of analysis for welding mechanics based on the FEM.

In 1971, Ueda and Yamakawa succeeded in an analysis of the thermal elastic-plastic behavior of butt joints during welding and published a paper on this pioneering work [1]. Since then, the authors and their colleagues have continued to analyze various types of welded joints including multipass joints of very thick plates. With these efforts, the group has established a simulation method for thermal elastic-plastic behavior of welded joints.

Parallel to these analyses they intended to examine the accuracy of the analysis compared to residual stresses measured on experimental models. During this process, they paid special attention to the source of residual stress, called inherent strain, and developed a very efficient method (inherent strain method) for predicting welding residual stress and deformation [2]. Additionally, they presented a new rational method for measuring three-dimensional residual stress in thick welded joints, utilizing the special features of inherent strain. These research results were compiled and published in *Computational Welding Mechanics* by the JWRI of Osaka University in 1999 [3].

Recently, the safety requirements of welded structures have become increasingly severe. To meet these requirements, engineers engaged in structural design and quality control need to anticipate welding deformation and residual stresses with higher accuracy. This gives engineers more opportunities to conduct computational analysis using software. To use the software and evaluate the output, engineers need a basic understanding of welding mechanics and experience using the software.

This book meets this demand by providing:

1. Basic theories and analysis procedures using a simple three-bar model.
2. A software package for performing basic analysis.
3. Examples of strategic methods and procedures to solve various welding-related problems encountered in the process of construction.
4. Appendices with databases of welding residual stresses, temperature-dependent material properties, etc.

In Chapters 1 and 2, using a simple three-bar model, the mechanism of production of residual stress and deformation during welding, the fundamental theory of measurement of residual stress, and the procedure of prediction of residual stress and deformation are presented. In Chapters 3 and 4, the basic theories of elasticity and plasticity, heat conduction, and the FEM are also explained using this model.

Chapters 5 and 6 present the documentation for the included software on the companion website in a Q&A format and illustrate how to use the programs through several examples. Chapter 7 presents suggestions for how to approach manufacturing problems relevant to welding mechanics through many examples of analysis based on the knowledge and accumulated experience of the authors.

Further, Appendix A offers data on welding residual stresses of analyses and experiments of various types of welding joints. Appendices B and C, found on the companion website at booksite.elsevier.com/9780123948045, present temperature-dependent mechanical properties of various materials used in the actual analysis and three-dimensional theory of thermal elastic-plastic analysis.

This book has been written for the practicing engineers who intend to utilize computational analysis and prediction of residual stress and deformation in their practical work. The authors expect that the readers will comprehend the capability of the computational methods of welding mechanics through using the software and accumulating new experience.

The authors believe that efforts of the readers should contribute to developing their own modern engineering system for manufacturing based on computational analysis and prediction.

The authors wish to thank their colleagues who were engaged in many research projects, of which results are referred to at many places in this book. Special thanks are extended to Elsevier's Science and Technology Editorial and Book Production team, for the support to publish this book.

Osaka
August 2011

REFERENCES

[1] Ueda Y, Yamakawa T. Analysis of thermal elastic-plastic stress and strain during welding by finite element method. Trans Japan Welding Soc 1971;2(2):90–100.

[2] Ueda Y. Prediction and measuring methods of two- and three-dimensional welding residual stresses by using inherent strain as a parameter. In Modeling in Welding, Hot Powder Forming, and Casting. Materials Park, OH: American Society for Metals, International; 1997.

[3] Ueda Y. Computational Welding Mechanics (in commemoration of retirement from Osaka University). The Joining and Welding Research Institute (JWRI) of Osaka University; 1999.

List of Symbols

A	Area, cross-sectional area
A_0	Cross-sectional area
a	Length of rectangle heat source
awsd.exe	Simple postprocessing program
B	Breadth, width
$[B]$	Strain-displacement matrix
B_F	Half breadth of flange
B_w	Depth of web
b	Width of rectangle heat source
b	Half width of inherent strain zone
b_0	Half width of inherent strain zone in case of infinitive plate width
$2b_0^{*2D}$	Width of inherent strain zone in two dimensions
$2b_0^{*3D}$	Width of inherent strain zone in three dimensions
C_0	Radiation constant for black body (Stefan-Boltzmann constant)
$[C]$	Heat capacity matrix
c	Specific heat
$[D]$	Elasticity matrix
D^{strain}, $[D^{strain}]$	
D^{temp}, $[D^{temp}]$	
D^{time}, $[D^{time}]$	
E	Young's modulus
E^p	Elastic-plastic modulus
e	Base of natural logarithm
F	Force, load
F^*	Sum of external force and thermal load
F^T	Tendon force
H'	Coefficient of work hardening
$[H^*]$	Elastic response matrix
$[H^*]^T$	Transposed matrix of $[H^*]$
h	Thickness, depth of rectangle heat source
heat2d.exe	Welding heat conduction FEM program
heat2d.inp	Input file for welding heat conduction FEM program
I	Welding current
inhs2d.exe	Inherent strain FEM program

inhs2d.inp	Input file for inherent strain FEM program
$[K]$	Stiffness matrix of whole structure
	Heat conduction matrix of whole structure
$[K]_e$	Stiffness matrix of element
$[K]^{-1}$	Inverse matrix of $[K]$
L	Length
ΔL	Longitudinal expansion in welding direction
ΔL^*	Inherent displacement, inherent deformation
m	Number of measured elastic strains, number of strain gages
P	Electric power
Q	Net heat input per unit length of weld
$\dot{Q}_A, \dot{Q}_B$	Time rate of heat quantity
q	Number of unknown effective inherent strains
q_A, q_B	Heat flow rate
$\dot{q}_A$	Surface heat flux
$\dot{q}_c$	Heat flows through the metal with unit cross-section in unit time
$\dot{q}_t$	Heat flow rate by convection
$\dot{q}_t$	Heat transfer per unit area and unit time
$\dot{q}_V$	Heat generation rate per volume
S	Sum of squares of the residuals ν_i
S_T	Transverse inherent displacement
ΔS	Transverse shrinkage
s	Cross-sectional area
$\hat{s}$	Unbiased estimate of measurement variance
T	Temperature
T_0	Initial temperature
T_{av}	Average temperature increase
T_m	Mechanical melting temperature
T_{max}	Maximum temperature
T_{melt}	Melting temperature
T_Y	Yield temperature
$\dot{T}(t)$	Rate of temperature change
$\dot{T}(\text{TC1}, \text{TC2})$	Average rate of temperature change between TC1 and TC2
ΔT	Temperature increment
t	Time
t_w	Heating time
Δt	Time increment
tepc2d.exe	Thermal elastic-plastic creep FEM program
tepc2d.inp	Input file for thermal elastic-plastic creep FEM program
U	Welding voltage
u	Displacement
u_Y	Displacement when temperature increment reaches T_Y

Δu	Displacement increment
v	Welding speed
x, y, z	Coordinates
DY	Displacement in y-direction
y_H	Half width of region where temperature reaches mechanical melting point
α	Instantaneous linear thermal expansion coefficient
α_0	Average value of linear thermal expansion coefficient
β	Equivalent heat transfer coefficient including both convection and radiation
β_c	Heat convection coefficient
$\ominus$	Circumferential direction of pipe
ε	Emissivity of the material
ε	Total strain
$\varepsilon_a, \varepsilon_b, \varepsilon_c$	Strain in bar **a**, bar **b**, and bar **c**, respectively
ε^c	Creep strain
ε^e	Elastic strain
ε^p	Plastic strain
ε^T	Thermal strain
ε_{YB}	Yield strain of base material
ε_{YW}	Yield strain of weld metal
$_m\varepsilon^e$	Measured elastic strain
$\varepsilon_x^*, \varepsilon_y^*, \varepsilon_z^*, \gamma_{yz}^*$	Components of inherent strain
$\hat{\varepsilon}^*$	Most provable value of effective inherent strain
$\overline{\varepsilon}_x^*$	Maximum value of inherent strain
$\overline{\varepsilon}_{x0}^*$	Maximum value of inherent strain in case of infinitive plate width
$\dot{\varepsilon}^c$	Creep strain rate
$\dot{\overline{\varepsilon}}^c$	Equivalent creep strain rate
$\Delta\varepsilon$	Strain increment
$\Delta\varepsilon_1^P$	Plastic strain increment during heating process
$\Delta\varepsilon_2^P$	Plastic strain increment during cooling process
$\Delta_m\varepsilon^e$	Observation errors
η	Efficiency of heat input
κ	Thermal diffusivity
λ	Thermal conductivity
ν	Poisson's ratio
$\{v\}$	Residuals
$\prod$	Total potential energy
σ	Stress
$\sigma_x, \sigma_y, \tau_{xy}$	Components of stress
σ_Y	Yield stress
$_m\sigma$	Measured stress

$\bar{\sigma}$	Equivalent stress
$\hat{\sigma}$	Most provable value of residual stress
$\Delta\sigma$	Stress increment
ρ	Density
***.post	Output file for postprocessor
AWS	American Welding Society
JSNAOE (SNAJ)	Society of Naval Architects and Ocean Engineering, Japan
JWS	Japan Welding Society
JWRI	Joining and Welding Research Institute, (former Welding Research Institute) Osaka University, Japan

Chapter 1

Introduction to Welding Mechanics

1.1 BASIC CONCEPTS OF WELDING AND WELDING MECHANICS

Fusion welding always produces thermal stress, deformation, and residual stress in products. The mechanism of production of stress and deformation and their influences upon the performance of welded structures are matters of keen concern for engineers. The theory of welding mechanics [1,2] presents the theoretical framework for understanding these consequences. In classic welding mechanics, theories are based on an empirical formula and using simplified models. However, recently, thermal elastic-plastic analysis [3–6] by the finite element method (FEM) has become popular and has been applied to simulate the mechanical behavior of metal during welding.

In practice, general structures are fabricated by a series of production processes such as cutting, bending, joining, straightening, residual stress relieving, etc. (as shown in Table 1.1). These processes may be classified into the following three groups: utilization of thermal process, mechanical device, and chemical reaction.

Welding belongs to the first group, so-called fusion joining (welding), that is, to join components by melting a small portion of the metal.

The energy sources for melting metal are gas, arc, plasma, electron beam, laser, etc. These are selected according to the type of structure and the efficiency of the manufacturing. For example, since a heavy pressure vessel is composed of many thick plates, a highly efficient heat source such as electron beam welding, which provides a concentrated large amount of heat input, is often adopted. For large structures such as ships, bridges, etc., arc welding is widely used because its machine is cheap and can be handled easily.

Generally, fusion welding is conducted by moving a heat source along the weld line. The joining process by any type of fusion welding is almost always the same and includes the heating process, melting process, and solidification process no matter what kind of welding heat source is employed. Therefore,

TABLE 1.1 Manufacturing Processes for Welded Structures

	Cutting	Forming	Joining	Stress Improvement	Straightening
Thermal processes	Thermal cutting (gas, plasma, laser)	Line heating (gas, high-frequency induction, laser)	Welding (gas, plasma, electron beam, laser, resistance welding, thermite welding); friction stir welding	Stress relief annealing; low-temperature stress relief	Line heating (gas, high-frequency induction, laser)
Mechanical processes	Saw; shear	Press; roller bending	Rivet, crimping	Shot peening	Press; hammer peening
Chemical processes	–	–	Bonding	–	–

arc welding is selected as an example, and the basic concepts of computational welding mechanics are presented here.

Many different types of welded joints, such as butt joints, fillet joints, etc., as illustrated in Fig. 1.1, are used according to the shape characteristics of the structural components. The groove type also differs according to plate thickness to obtain sufficient penetration, as shown in Fig. 1.2 [7].

In arc welding, the surface of the groove of the base metal is heated and molten by arc heating. Simultaneously, the electrode is also molten and fills into the groove. After cooling down, the joint is formed. Figure 1.3 is a schematic illustration of the electric power source and the welding arc. The type of welding power source, such as direct current, alternating current, and pulse current, must be selected to fit the type of welded joint.

In arc welding, the heat input per unit weld length is expressed by

$$Q = \eta I U / v. \tag{1.1.1}$$

In this equation, the current of the direct-current source is denoted by I, the voltage between the electrode and the base plate by U, and the traveling speed of the electrode by v. η is the efficiency of the heat input, which takes into account loss of electric energy by radiation, convection, etc. For example, in submerged arc welding, a high efficiency such as $\eta = 0.9$–0.99 is expected, and $\eta = 0.3$~0.6 for low efficiency of TIG (tungsten inert-gas) welding [5].

When the heat source of welding travels along the weld line, the temperature distribution in the base plate changes as a function of time elapsed.

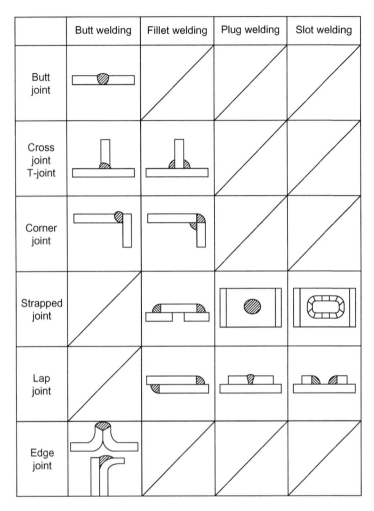

FIG. 1.1 Types of welded joints.

For example, when a point heat source moves on the surface of a thick plate, the temperature distributions are calculated as indicated in Fig. 1.4 [8]. Beneath the source, the temperature is the highest and gradually decreases from the heat source toward the starting point of welding. The temperature distribution in the vertical plane against the weld line indicates a concentric circle. As a result, the portion where the temperature is above the melting point is regarded as the weld metal, and the surrounding area, in the case of steel, which is heated above the A_{c1} transformation temperature (723°C), forms the heat-affected zone (HAZ), as illustrated in Fig. 1.5. The HAZ exhibits different mechanical properties in hardness and ductility, which influence the strength and reliability of the welded joint.

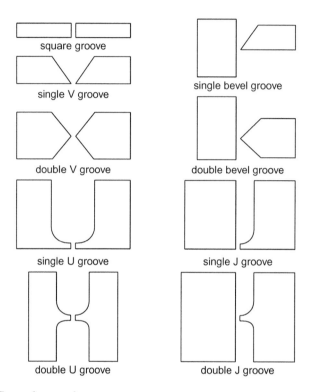

FIG. 1.2 Types of groove shapes.

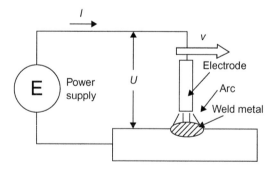

FIG. 1.3 Schematic illustration of electric power and welding arc.

Problems related to welding may be classified into three groups as shown in Table 1.2:

1. **Metallurgical changes:** induced through thermal histories.
2. **Welding process:** supplies heat input for melting and solidification of the metal.

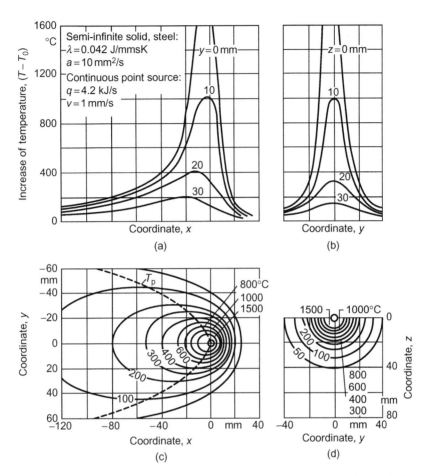

FIG. 1.4 Calculated temperature distributions on the surface of a thick plate using a point heat source [8].

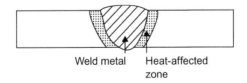

FIG. 1.5 Section of a welded joint.

3. Mechanics: welding deformation and residual stress as a consequence of the welding process.

When problems with welding deformation and residual stress are addressed, Table 1.2 suggests that it is essential to take into consideration the influences of

TABLE 1.2 Relationships Among the Welding Process, Material, and Mechanics

	Influential Factors	Consequence
Welding process	Welding method, current, voltage, welding speed, groove	Heat input, temperature distribution, heat efficiency, shapes of reinforcement and penetration
Material	Chemical composition, temperature, thermal history, cooling rate	Micro-structure, phase transformation, thermal and mechanical properties, hardness, strength, toughness
Mechanics	Temperature distribution and history, thermal and mechanical properties, shape and size of work, constraint	Residual stress, distortion, welding crack

the welding process, which affect shape and depth of penetration, and metallurgical changes, which affect thermal expansion and shrinkage through phase transformation, etc.

Welding deformation and residual stress are produced by plastic shrinkage that is induced as a result of thermal expansion and shrinkage in the welded portion during thermal processes. Figure 1.6 illustrates basic patterns of welding deformation in a butt joint. They are transverse and longitudinal shrinkages, and rotative and angular distortions. Figure 1.7 represents typical patterns of residual stress distributions in a butt-welded joint. Figures 1.7 (a) and (b) show longitudinal residual stress σ_x and transverse stress σ_y along the weld line and in the transverse section, respectively. The maximum magnitude of longitudinal stress is

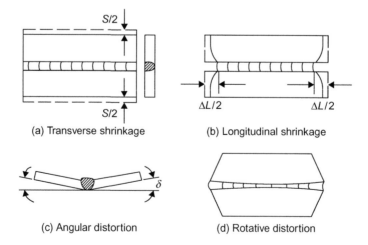

(a) Transverse shrinkage　　(b) Longitudinal shrinkage

(c) Angular distortion　　(d) Rotative distortion

FIG. 1.6　Basic types of welding deformation.

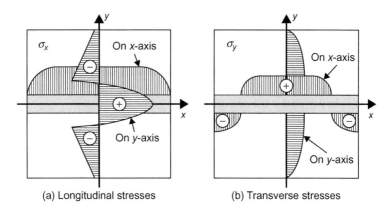

(a) Longitudinal stresses   (b) Transverse stresses

FIG. 1.7 Typical distribution patterns of welding residual stresses in a butt-welded joint.

TABLE 1.3 Welding-induced Problems and Countermeasures

	Problems	Countermeasures
Welding residual stress	Brittle fracture, fatigue, stress corrosion crack (SCC), buckling	Heat input, welding sequence, post-weld heat treatment (PWHT)
Welding distortion	Geometrical inaccuracy, anti-automation	Prediction, offset, constraint, clamping, straightening
Cracking	Hot/cold cracking	Weld condition, material selection, pre-/postheating

approximately equal to the yield stress. Unless the combination of base and filler metals and the welding condition are chosen properly, weld cracks may occur in the welded portion. The residual stress, distortion, and weld cracks may deteriorate the strength of the welded structure and lower the reliability of the product as described in Table 1.3. Then, it is suggested to take appropriate measures based on prediction and/or simulation. For example, stress relief annealing [9] can be applied to reduce the residual stress and improve metallurgical properties. In welded distortion, prestrain may be given to compensate for the resulting distortion.

To prepare for the consequences of welding, prediction or estimation of residual stress and distortion is necessary. One of the most powerful methods is to use a computer to simulate behavior during welding.

To meet this need, this book provides the following:

1. The basic theory of thermal elastic-plastic analysis, rewritten to apply the FEM.
2. Demonstration of practical steps for the analysis of basic examples using software from the companion website.

3. Presentation of several applications of simulation to practical problems, explanations of the mechanism of the production of residual stress and distortion, and suggestions for solutions to the problems.

As described so far, the main goal of this book is to analyze mechanical behavior during welding by the FEM using a computer. This method of analysis is called computational welding mechanics.

The direct method of computational simulation is a main theme in this book. However, another important theme is also introduced in the book, the concept of inherent strain, which is the source of residual stress and distortion. With the help of this concept, the results of our experiments and computer simulation can be analyzed, and the mechanism of the production can be explained clearly. Furthermore, as this concept is based on the theory of elasticity, it can be applied to estimate residual stress and distortion very efficiently.

In this chapter, using a simple three-bar model, the production process of residual stress and distortion is demonstrated, and the mechanism is explained by the concept of inherent strain. Details of the inherent strain method is presented in Chapter 2.

1.2 PROCESS OF THE PRODUCTION OF RESIDUAL STRESS AND INHERENT STRAIN

1.2.1 Illustrative Model

1.2.1.1 Temperature Distribution in a Welded Joint

During welding, a large amount of heat input is supplied in a limited portion of a joint. This portion and its vicinity are molten so that the separate portions of the joint are connected. The temperature distribution in the joint is not uniform and changes with time. Figure 1.4 demonstrates the temperature distribution of a butt joint under a moving heat source, shown in isothermal curves. The temperature is highest around the heat source, and the shape of the isothermal curves is almost a circle. Apart from the heat source, isothermal curves become elliptic. When the joint is sufficiently long, the same temperature distribution is moving with the moving heat source except near the starting and finishing points of welding. Then, this condition is regarded as quasistatic.

It is also observed that the slope of temperature gradient around the heat source is very steep in any direction, and apart from the heat source along the welding direction, it becomes gentle in its direction, but remains steep in the transverse direction.

1.2.1.2 Temperature Distribution and Mechanical Constraint

Uneven temperature distribution as observed above, causes different magnitude of thermal expansion in the joint. In order to adjust the difference in the joint, elastic and/or plastic strains are produced and consequently result in stresses and distortion.

First, as the temperature gradient is very sharp, we may classify the temperature field into two regions: very high temperature near the heat source and low temperature in the adjacent surrounding area. It is natural that the thermal expansion of the high-temperature zone is under the restraint of the surrounding area. In a long butt joint, the same temperature field is moving under the quasistatic condition as stated above. As a consequence of this condition, the transverse section of the joint is supposed to keep plane during welding in most parts of the joint except the start and end. This mechanical behavior is caused by the strong longitudinal constraint against thermal expansion, but the transverse constraint is so low as to be presumed free.

1.2.1.3 Idealized Model

To simulate the mechanical behavior of a butt-welded joint in the longitudinal direction, a simple model shown in Fig. 1.8, which consists of three bars, is proposed. The center bar, C, corresponds to the high-temperature region of the welded metal and its vicinity, and the two side bars are the remaining part of the joint, of which volume is large and temperature is low. The longitudinal gradient of temperature of the joint is so gentle that the temperatures of the three bars are assumed to be uniform along the length. When the volume of the side bars is so large and their temperature changes so little, the center bar is regarded as fixed at both ends.

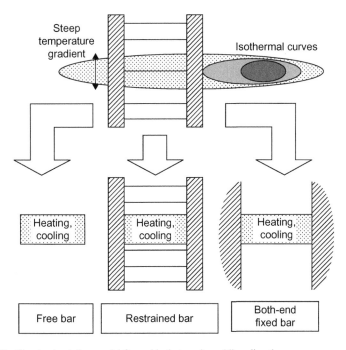

FIG. 1.8 Simple simulation model for residual stress in welding direction.

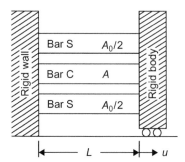

FIG. 1.9   Model for analysis.

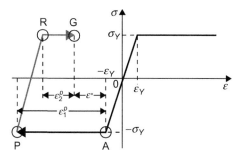

FIG. 1.10   Assumed stress and strain curve.

The proposed model is represented again in Fig. 1.9. The three bars are bar C and two bar S, of which the length is the same, and the cross-sectional areas are A and $A_0/2$, respectively. The model is fixed at one end and connected to the movable rigid body at the other end. For simplicity, the material is assumed to be elastic perfectly plastic, with Young's modulus E and yield stress σ_Y, as shown in Fig. 1.10. It may be noted that the stress in the elastic range $(-\sigma_Y \leq \sigma \leq \sigma_Y)$ is proportional to the strain and no stress increment is assumed in the plastic range. In the same figure, when the stress is unloaded from P in the compressive plastic range, the unloaded process is elastic to reach R at the tensile yield stress, and no stress increment is observed, but plastic flow. The material is also assumed not to be temperature dependent so that the thermal expansion coefficient α is constant. The base temperature is set at 0°C.

1.2.1.4 Thermal Elastic-Plastic Analysis

In this section, we perform thermal elastic-plastic analysis of the model under several heating and cooling conditions to observe the process of the production of residual stress and deformation. As a consequence of the analysis, it is found that residual stress is produced by plastic strain induced under the thermal history, which is the source of residual stress, called inherent strain. Accordingly, we can calculate residual stress if we impose the inherent strain on the stress-free joint and perform standard elastic analysis. This indicates that we can predict residual stress simply by standard elastic analysis, if we have data on inherent strain to the weld joint concerned. Section 1.3 presents the inverse method of standard elastic analysis, by which the inherent strains can be obtained from the results of experiments and thermal elastic-plastic analysis. Applying the inverse method of analysis, we can accumulate data on inherent strains to construct a database.

We carry out the analysis on the model under the following three different restraint conditions as shown in Fig. 1.11, providing heating and cooling processes to bar C:

1. Restraint condition ①: Bar C is separated from the other bars and becomes a free bar. Then, it can be elongated or shortened freely under any change of temperature, but neither thermal nor residual stress is induced in bar C. The same phenomenon is observed when the three bars are simultaneously subjected to the same temperature change.

2. Restraint condition ②: Bar C is fixed at both ends. When the cross-sectional area of bar S tends to infinitively large, the restraining condition of bar C approaches to be fixed at both ends. This condition may be referred to a butt-welded joint with a very wide plate in actual case. In this case, thermal

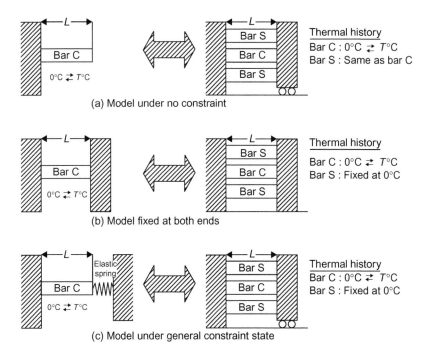

FIG. 1.11 Three types of restraint conditions for three-bar model.

deformation of bar C is completely resisted by bar S through the movable rigid body. Depending on the heating temperature of bar C, plastic strain may be produced, which becomes the source of residual stress.

3. Restraint condition ③: This is the most general case of the model where bar C plays the role of one of the structural members. Thermal expansion and shrinkage of bar C is resisted by bar S through the movable rigid body. Under a given thermal history, plastic strain may be induced in bar C, which causes residual stress.

Here, we perform thermal elastic-plastic analysis on the above three mechanical conditions. The analysis should be carried out so as to satisfy the following basic conditions:

1. Displacement and strain relation
2. Stress and strain relation
3. Equilibrium condition
4. Boundary condition

1.2.2 Heating of Free Bar C (Restraint condition ①)

When all three bars in Fig. 1.9 are heated to the same temperature simultaneously, they elongate as if they were free bars, as shown in Fig. 1.12.

When bar C is heated to $T°C$ from $0°C$, it elongates by u, which is given by

$$u = \alpha T L. \tag{1.2.1}$$

In this condition, the elongation is free from any constraints, so that no stress is induced. Next, the bar is cooled to $0°C$, and the length of bar C is shrunken so that the elongation u vanishes. Accordingly, neither displacement nor stress is observed. The temperature and stress relation during this process is represented in Fig. 1.13.

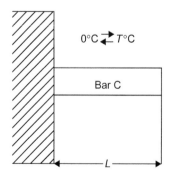

FIG. 1.12 Free bar.

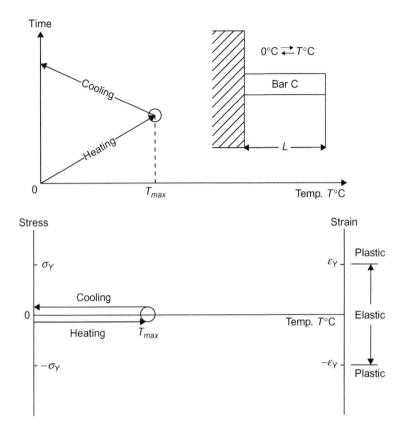

FIG. 1.13 Thermal and stress history of free bar.

1.2.3 Heating of Bar C Fixed at Both Ends (Restraint condition ②)

Bar C in Fig. 1.14 is fixed at both ends, which can be realized in the model if the movable rigid body is clamped.

The mechanical behavior of bar C is analyzed under the condition that bar C is heated up to the maximum temperature T_{max}°C from 0°C and cooled down to 0°C. The elastic-plastic behavior is different depending on the magnitude of the maximum temperature T_{max}, which may cause plastic strain and residual stress if T_{max} exceeds yield temperature T_Y, which is defined as the temperature that causes the yielding in a bar fixed at both ends either by heating or cooling, that is,

$$T_Y = \frac{\sigma_Y}{\alpha E}. \tag{1.2.2}$$

This is theoretically derived in Eq. (1.2.5).

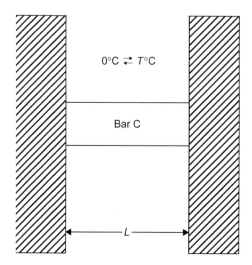

FIG. 1.14 Bar fixed at both ends.

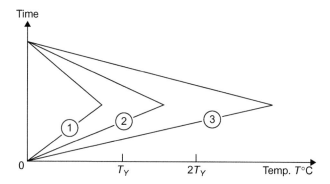

FIG. 1.15 Thermal histories with three different maximum temperatures.

As illustrated in Fig. 1.15, the heating and cooling processes are classified into the following three cases with respect to yield temperature T_Y:

1. Low-temperature heating (process ①, $0°C \rightarrow T_{max} \leq T_Y \rightarrow 0°C$)
2. Medium-temperature heating (process ②, $0°C \rightarrow T_Y \leq T_{max} \leq 2T_Y \rightarrow 0°C$)
3. High-temperature heating (process ③, $0°C \rightarrow 2T_Y \leq T_{max} \rightarrow 0°C$)

During process ①, neither plastic strain nor residual stress is induced. During process ②, plastic strain is produced at the heating stage, which causes residual stress when the bar is cooled down. During process ③, plastic strains are produced both at the heating and cooling stages, and the sum of these plastic strains is the source of residual stress.

1.2.3.1 Low-Temperature Heating
(Process ①, $0°C \rightarrow T_{max} \leq T_Y \rightarrow 0°C$, Fig. 1.16)

Heating Stage (Heating from $T = 0°C \Rightarrow T_{max} \leq T_Y$)

We heat bar C from $0°C$ to $T_{max} \leq T_Y$. If bar C is free, it elongates by αTL due to thermal expansion. However, as the bar is fixed at both ends, the elongation is completely suppressed by the rigid wall and results in compressive strain. The stress point of bar C starts from the origin to point A in Fig. 1.16. This mechanical behavior is analyzed in the following procedure.

In general, total strain is defined as the sum of each component. In this case, as bar C is in the elastic range, the total strain ε is composed of elastic strain ε^e and thermal strain $\varepsilon^T = \alpha T$,

$$\varepsilon = \varepsilon^e + \varepsilon^T. \tag{1.2.3}$$

As the bar is clamped at both ends and its elongation is restricted, the total strain should be $\varepsilon = 0$. This leads to the following result.

$$\varepsilon^e = -\alpha^T \tag{1.2.4}$$

$$\sigma = E\varepsilon^e = -E\alpha T.$$

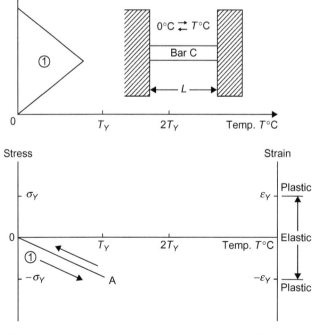

FIG. 1.16 Histories of thermal stress and strain of bar fixed at both ends. (Low-temperature heating, process ①.)

When the bar is heated further, the thermal stress of the bar increases to reach the yield stress that is at A in Fig. 1.16. Denoting this temperature by T_Y, this is given by

$$-\alpha T_Y E = -\sigma_Y, \quad T_Y = \frac{\sigma_Y}{\alpha E}. \tag{1.2.5}$$

Cooling Stage (Cooling from $T_{max} \leq T_Y$ to 0°C)

Cooling down bar C from $T_{max} \leq T_Y$ to 0°C, its stress point returns to the origin of Fig. 1.16. The behavior of the bar at this stage is elastic so that no residual stress is produced.

1.2.3.2 Medium-Temperature Heating
(*Process ②, 0°C → $T_Y \leq T_{max} \leq 2T_Y$ → 0°C, Fig. 1.17*)

Heating Stage

We heat up bar C from 0°C to $T_Y \leq T_{max} \leq 2T_Y$. At the temperature T_Y, the compressive stress induced in the bar reaches the yield stress σ_Y. The stress point moves from 0 to A in Fig. 1.17. Heating the bar further above T_Y by an increase

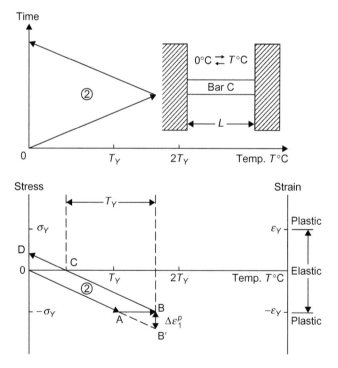

FIG. 1.17 Histories of thermal stress and strain of a bar fixed at both ends. (Medium-temperature heating, process ②.)

of temperature $\Delta T = (T_{max} - T_Y)$ induces additional thermal strain $\alpha \Delta T$ in the bar. As the bar is fixed at both ends, this additional strain $\alpha \Delta T$ is compressed as compressive plastic strain due to plastic flow, and the strain reaches B' in the same figure. As the stress is at the yield stress, no further stress increase is expected. In this case, an increment of total strain $\Delta \varepsilon$ is composed of increments of elastic strain, thermal strain, and plastic strain, which are denoted by $\Delta \varepsilon^e$, $\Delta \varepsilon^p$, and $\Delta \varepsilon^T = \alpha \Delta T$, respectively, as

$$\Delta \varepsilon = \Delta \varepsilon^e + \Delta \varepsilon^p + \Delta \varepsilon^T. \tag{1.2.6}$$

As the bar is clamped at both ends, the total strain should be kept at zero. The stress has reached the yield stress and no further increase of stress (elastic strain) is allowed. Then, Eq. (1.2.6) is rewritten as

$$0 = \Delta \varepsilon^p + \Delta \varepsilon^T$$
$$\Delta \varepsilon^p = -\Delta \varepsilon^T = -\alpha(T_{max} - T_Y) \equiv \Delta \varepsilon_1^p. \tag{1.2.7}$$

This implies that the increase of thermal strain becomes that of the compressive plastic strain, and the bar is shortened by the compressive plastic strain. At this stage, the stress point moves to B in the figure.

Cooling Stages

① **Stage 1** (cooling from $T = T_{max}$ by T_Y, i.e., from stress point B to C, Fig. 1.17).

Cooling the bar by T_Y from T_{max}, the temperature reaches $T = T_{max} - T_Y$, and the shrinkage strain increment $\Delta \varepsilon^T$ is expressed by

$$\Delta \varepsilon^T = -\alpha T_Y.$$

The bar is in the unloading stage and the behavior is elastic. The stress point moves from B to C in process ② in Fig. 1.17. As the bar is fixed at both ends, no increments in total strain and plastic strain are induced in this process. Then, the following equation is deduced, which leads to the stress change:

$$0 = \Delta \varepsilon = \Delta \varepsilon^e + \Delta \varepsilon^T = \Delta \varepsilon^e - \alpha T_Y$$
$$\Delta \varepsilon^e = \alpha T_Y \tag{1.2.8}$$
$$\therefore \Delta \sigma = E(\alpha T) = \sigma_Y.$$

Prior to this process, the stress is $-\sigma_Y$, and now that σ_Y is added, the resulting stress is zero,

$$\sigma = -\sigma_Y + \sigma_Y = 0, \tag{1.2.9}$$

and the stress point moves to C.

② **Stage 2** (cooling from $T = (T_{max} - T_Y)$ to $T = 0°C$, i.e., from stress point C to D, Fig. 1.17).

Bar C is cooled down to $T = 0°C$ by $T_{max} - T_Y$. The shrinkage strain $-\alpha(T_{max} - T_Y)$ induced in this stage is constrained by the rigid wall, and the strain becomes tensile elastic strain, which results in tensile stress. The stress point reaches D in the figure, and the final stress is given as

$$0 = \Delta\varepsilon = \Delta\varepsilon^e + \Delta\varepsilon^T = \Delta\varepsilon^e - \alpha(T_{max} - T_Y)$$
$$\Delta\varepsilon^e = \alpha(T_{max} - T_Y) \tag{1.2.10}$$
$$\sigma = \alpha E(T_{max} - T_Y).$$

By this stage of cooling, the temperature becomes zero and no thermal strain exists. Therefore, the residual stress is produced by the plastic strain, $\Delta\varepsilon^p = -\alpha(T_{max} - T_Y)$, which is induced at the heating stage. In this case, this plastic strain is the source of residual stress, which is inherent strain ε^*,

$$\varepsilon^* = -\alpha(T_{max} - T_Y). \tag{1.2.11.A}$$

During this process, the maximum temperature is maintained below $T_{max} = 2T_Y$, and the plastic strain is produced in the heating stage. Here, $T_{max} = 2T_Y$ is called the limit temperature of the medium-temperature heating process. If the bar becomes free from the rigid wall, it becomes shorter by ΔL^*, which is called inherent displacement. This is given by integration of the plastic strains as

$$\Delta L^* = \varepsilon^* L = -\alpha(T_{max} - T_Y)L. \tag{1.2.11.B}$$

1.2.3.3 High-Temperature Heating
(Process ③, $0°C \rightarrow 2T_Y \leq T_{max} \rightarrow 0°C$, Fig. 1.18)

Heating Stage (Heating from $T = 0°C$ to $2T_Y \leq T_{max}$, i.e., Stress Point 0 to P, Fig. 1.18)

In this case, the maximum temperature of bar C reaches above $2T_Y \leq T_{max}$ and the thermal strain reaches P′ in the figure. As the bar is in the plastic range, the stress stays at σ_Y and reaches point P in the figure, and the thermal strain induced above T_Y,

$$\Delta\varepsilon^T = \alpha\Delta T = \alpha(T_{max} - T_Y),$$

is suppressed to turn into compressive plastic strain,

$$\Delta\varepsilon^p = -\Delta\varepsilon^T = -\alpha(T_{max} - T_Y) \equiv \Delta\varepsilon_1^p.$$

Cooling Stages

① **Stage 1** (cooling from $T_{max} > 2T_Y$ by T_Y, i.e., stress point P to Q, Fig. 1.18).
At this cooling stage, the shrinkage strain of the bar is $-\alpha T_Y$, which reduces the stress by σ_Y, and the stress in the bar vanishes. The stress point moves from P to Q.
② **Stage 2** (cooling further by T_Y to $T = T_{max} - 2T_Y$).
The shrinkage strain by this cooling is $-\alpha T_Y$, which is resisted by the wall and turns into tensile yield stress σ_Y. The stress point reaches R in Fig. 1.18.

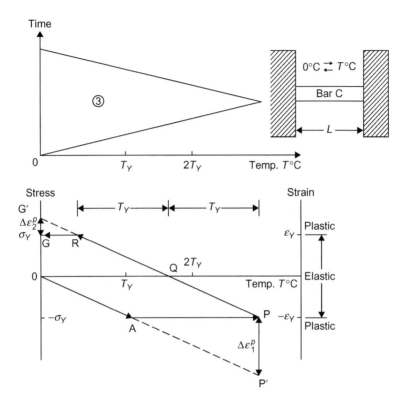

FIG. 1.18 Histories of thermal stress and strain of a bar fixed at both ends. (High-temperature heating, process ③.)

③ **Stage 3** (cooling to $T = 0°C$, i.e., stress point R to G, Fig. 1.18).

At this stage, the shrinkage strain is $-\alpha(T_{max} - 2T_Y)$, which is resisted completely. As the stress is at the tensile yielding, no further increase of stress occurs, and the shrinkage strain turns into tensile plastic strain. Stress point R shifts to G, and the plastic strain increment is given by

$$0 = \Delta \varepsilon = \Delta \varepsilon^p + \Delta \varepsilon^T = \Delta \varepsilon^p - \alpha(T_{max} - 2T_Y)$$
$$\Delta \varepsilon^p = \alpha(T_{max} - 2T_Y) \equiv \Delta \varepsilon_2^p. \tag{1.2.12}$$

During this process of high-temperature heating, plastic strain is produced in two stages: at the heating stage, $\Delta \varepsilon^p = -\alpha(T_{max} - T_Y) \equiv \Delta \varepsilon_1^p$, and at stage 3 of the cooling stage, $\Delta \varepsilon^p = \alpha(T_{max} - 2T_Y) \equiv \Delta \varepsilon_2^p$. The former is compressive and the latter is tensile. Then, the sum of these two results is residual plastic strain, that is,

$$\Delta \varepsilon^p = -\alpha(T_{max} - T_Y) + \alpha(T_{max} - 2T_Y) = -\alpha T_Y. \tag{1.2.13}$$

This plastic strain plays a role as the source of residual stress, that is, inherent strain ε^*,

$$\varepsilon^* = \Delta\varepsilon_1^p + \Delta\varepsilon_2^p = -\alpha T_Y. \qquad (1.2.14.A)$$

The corresponding inherent deformation is given by

$$\Delta L^* = \varepsilon^* L = -\alpha T_Y L. \qquad (1.2.14.B)$$

The preceding analysis demonstrates that the difference of the maximum temperature T_{max} at the heating stage affects the process of the production of the resulting plastic strain, which is the source of residual stress. As a consequence, the magnitude of residual stress is influenced. Correlations between these factors are illustrated in Fig. 1.19.

In the practical field of welding engineering, estimation of residual stresses with reasonable accuracy is usually required. As we learned in this section, residual stress can be calculated by standard elastic analysis imposing inherent strain on a stress-free specimen. This method is explained further in Section 1.3.

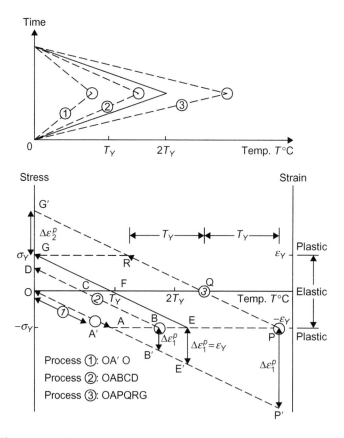

FIG. 1.19   Histories of temperature, stress, and strain of bar C under different thermal processes.

TABLE 1.4 Summary of Plastic Strain $\Delta\varepsilon^p$, Inherent Strain ε^*, Inherent Deformation ΔL^*, and Residual Stress σ of Bar C Under Three Different Heating Processes

Heating Process	Low Temp.	Medium Temp.	Limit Temp.	High Temp.
Maximum temp.: T_{max}	$T_{max} \leq T_Y$	$T_Y \leq T_{max} \leq 2T_Y$	$T_{max} \leq 2T_Y$	$2T_Y \leq T_{max}$
Plastic strain in heating process: $\Delta\varepsilon_1^p$	0	$-\alpha(T_{max} - T_Y)$	$-\varepsilon_Y$	$-\alpha(T_{max} - T_Y)$
Plastic strain in cooling process: $\Delta\varepsilon_2^p$	0	0	0	$\alpha(T_{max} - 2T_Y)$
Inherent strain: $\varepsilon^* = \Delta\varepsilon_1^p + \Delta\varepsilon_2^p$	0	$-\alpha(T_{max} - T_Y)$	$-\varepsilon_Y$	$-\alpha T_Y = -\varepsilon_Y$
Inherent deformation $\Delta L^* = \varepsilon^* L$	0	$-\alpha(T_{max} - T_Y)L$	$-\varepsilon_Y L$	$-\alpha T_Y L = -\varepsilon_Y L$
Residual stress: σ	0	$\alpha E(T_{max} - T_Y) \leq \sigma_Y$	σ_Y	σ_Y

Here, the correlations between plastic strain, inherent strain, and inherent deformation, which are produced under the three different thermal processes, are summarized in Table 1.4. When the base temperature is $T°C$ instead of $0°C$, the yield temperature T_Y should be modified to $T_Y = (\sigma_Y/\alpha_E) + T°C$.

In Section 1.2.4, where the movable rigid body is free to move, T_Y should be $T_Y = \dfrac{A + A_0}{A_0} \dfrac{\sigma_Y}{E\alpha}$ for the base temperature $0°C$ and $T_Y = \dfrac{A + A_0}{A_0} \dfrac{\sigma_Y}{E\alpha} + T°C$ for $T°C$.

1.2.4 Heating Bar C When It Is Connected to a Movable Rigid Body (Restraint condition ③)

Here, we analyze the mechanical behavior of the model in Section 1.2.1, of which the left rigid wall is fixed and the right one is movable in parallel to the left, as illustrated in Fig. 1.20. The heating and cooling processes are the same as those in Section 1.2.3, such as $T = 0°C \rightarrow T_{max}°C \rightarrow 0°C$.

In this case, the process of the production of residual stress is governed by the magnitude of maximum temperature T_{max}. Similar to Section 1.2.3, the analysis is performed for three different thermal processes as shown in Fig. 1.21. The resulting relationships among temperature, strain, and stress are represented by the same figures given in Section 1.2.3. In other words, the consequences of the three heating processes are illustrated in Figs. 1.16–1.18. With this in mind, we perform the analysis on the following three heating processes:

1. Low-temperature heating (process ①, $0°C \rightarrow T_{max} \leq T_Y \rightarrow 0°C$)
2. Medium-temperature heating (process ②, $0°C \rightarrow T_Y \leq T_{max} \leq 2T_Y \rightarrow 0°C$)
3. High-temperature heating (process ③, $0°C \rightarrow 2T_Y \leq T_{max} \rightarrow 0°C$)

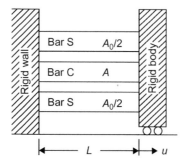

FIG. 1.20   Bar C with elastic restraint.

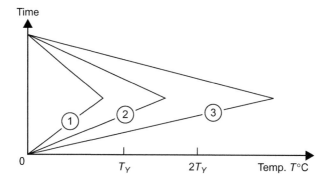

FIG. 1.21   Three different thermal processes with different temperatures of T_{max}.

Here, it should be noted that yield temperature T_Y in the above is specified by Eq. (1.2.21), which is different from that defined in Section 1.2.3.

1.2.4.1 Low-Temperature Heating
(Process ①, 0°C → $T_{max} \leq T_Y$ → 0°C, Fig. 1.16)

Heating Stage (Heating from $T = 0°C \Rightarrow T_{max} \leq T_Y$)

We heat bar C from 0°C to $T_{max} \leq T_Y$. If bar C is free, it elongates by αTL due to thermal expansion. However, as the bar is connected to the movable rigid body, the magnitude of thermal expansion is controlled by bar S through the wall. Consequently, compressive stress is induced in bar C and tensile stress in bar S. In this process, T_{max} is below T_Y and all bars behave elastically. The movable rigid body moves by u, and all bars are subjected to the same displacement. Total strain ε of the bars yields

$$\varepsilon = u/L. \tag{1.2.15}$$

In this process, thermal strain is observed only in bar C and not in bar S, and the total strain is expressed as

$$\left.\begin{array}{ll} \text{bar S:} & \varepsilon = \varepsilon^e \\ \text{bar C:} & \varepsilon = \varepsilon^e + \varepsilon^T \end{array}\right\}. \tag{1.2.16}$$

Therefore, strain and stress are shown as

$$\left.\begin{array}{lll} \text{bar S:} & \varepsilon^e = u/L, & \sigma = Eu/L \\ \text{bar C:} & \varepsilon^e = u/L - \alpha T, & \sigma = E(u/L - \alpha T) \end{array}\right\}. \tag{1.2.17}$$

The stress point of bar C shifts from 0 to A in Fig. 1.16. The strains and stresses are a function of u, which can be determined from the equilibrium condition of the three bars, as follows.

The total cross-sectional area of bar S is A_0, and the condition is written as

$$\begin{aligned} 0 &= A_0\,Eu/L + AE(u/L - \alpha T) \\ u &= A/(A + A_0) \cdot \alpha TL. \end{aligned} \tag{1.2.18}$$

Substituting this u into Eq. (1.2.17), the stress of each bar is given by

$$\left.\begin{array}{ll} \text{bar S:} & \sigma = E\dfrac{A}{A + A_0}\alpha T \\[2mm] \text{bar C:} & \sigma = -E\dfrac{A_0}{A + A_0}\alpha T \end{array}\right\}. \tag{1.2.19}$$

When the temperature increases to $T = T_Y$, the stress of bar C reaches the yield stress and the stress point moves to A in Fig. 1.16. At this point, the displacement of the wall being denoted by u_Y, T_Y is derived by imposing the yield condition to Eq. (1.2.17),

$$\begin{aligned} E\,(u_Y/L - \alpha T_Y) &= -\sigma_Y \\ T_Y &= \left(\dfrac{\sigma_Y}{E} + \dfrac{u_Y}{L}\right)\Big/\alpha. \end{aligned} \tag{1.2.20}$$

At the same time, T and u in Eq. (1.2.18) are rewritten as T_Y and u_Y. They are derived from Eqs. (1.2.18) and (1.2.20) as

$$T_Y = \frac{A + A_0}{A_0}\frac{\sigma_Y}{E\alpha} \tag{1.2.21}$$

$$u_Y = \frac{A}{A + A_0}\alpha\left(\frac{A + A_0}{A_0}\right)\frac{\sigma_Y L}{E\alpha} = \frac{A}{A_0}\frac{\sigma_Y L}{E}. \tag{1.2.22}$$

Here, attention should be paid to T_Y in Eq. (1.2.21), which is expressed by T_Y in Eq. (1.2.5), modified by coefficient $(A + A_0)/A_0$. This coefficient represents the degree of restraint due to bar S. Usually this is larger than 1 and approaches to 1 as the cross-sectional area of bar S tends to infinity, which is equivalent to the case where the movable rigid body is fixed.

Figure 1.16 represents the relationships between temperature, strain, and stress of bar C fixed at both ends, which was analyzed in Section 1.2.3. This can be used in the present case, noting that T_Y is modified by coefficient $(A + A_0)/A_0$, as indicated in Eq. (1.2.21).

Cooling Stage (Cooling from $T_{max} \leq T_Y$ to 0°C)

When bar C cools down from T_{max} to 0°C its stress point returns to the origin of Fig. 1.16. The behavior of the bar in this stage is elastic so that no residual stress is produced.

1.2.4.2 Medium-Temperature Heating
(Process ② 0°C → $T_Y \leq T_{max} \leq 2T_Y$ → 0°C, Fig. 1.17)

Heating Stage (Heating from $T = 0°C \Rightarrow T_Y \leq T_{max} \leq 2T_Y$)

Consider bar C heated from 0°C to $T_Y \leq T_{max} \leq 2T_Y$. At temperature T_Y, compressive stress induced in the bar reaches the yield stress, and the stress point moves from 0 to A in Fig. 1.17. When bar C is heated further above T_Y by $\Delta T = (T_{max} - T_Y)$ to T_{max}, the corresponding additional thermal strain $\alpha \Delta T$ is produced in the bar.

As the bar is yielded, no further increase of stress is allowed so that no force to push the movable rigid body is produced. Then the additional thermal strain remains in the bar as compressive plastic strain, and the strain reaches B′ in the same figure.

In this case, an increment of total strain $\Delta \varepsilon$ is composed of increments of thermal strain and plastic strain, but it should be zero. Therefore,

$$\Delta \varepsilon = \Delta \varepsilon^p + \Delta \varepsilon^T$$

$$0 = \Delta \varepsilon = \Delta \varepsilon^p + \Delta \varepsilon^T = \Delta \varepsilon^p + \alpha(T_{max} - T_Y)$$
$$\Delta \varepsilon^p = -\alpha(T_{max} - T_Y) \equiv \Delta \varepsilon_1^p. \tag{1.2.23}$$

As the stress does not change at this stage, the moveable rigid body stays at u_Y, which is produced at $T = T_Y$.

Cooling Stages

① **Stage 1** (cooling from $T = T_{max}$ by T_Y, i.e., from stress point B to C, Fig. 1.17).

Cooling the bar by T_Y from T_{max}, the temperature reaches $T = T_{max} - T_Y$, and the shrinkage strain increment $\Delta \varepsilon^T$ is

$$\Delta \varepsilon^T = -\alpha T_Y.$$

The bar is in the unloading stage, and the two bars S are also unloaded elastically. Accordingly, the rigid body displaces by Δu. A change of total strain $\Delta \varepsilon$ may be expressed in terms of Δu as

$$\left. \begin{array}{ll} \text{bar S:} & \Delta \varepsilon = \Delta u / L = \Delta \varepsilon^e \\ \text{bar C:} & \Delta \varepsilon = \Delta u / L = \Delta \varepsilon^e + \alpha T_Y \end{array} \right\}. \tag{1.2.24}$$

As $\Delta\varepsilon^e$ is an elastic strain increment, increments of elastic strain and stress, $\Delta\varepsilon^e$ and $\Delta\sigma$, are expressed as

$$\left.\begin{array}{ll} \text{bar S:} & \Delta\varepsilon^e = \Delta u/L \\ \text{bar C:} & \Delta\varepsilon^e = \Delta u/L - \alpha T_Y \end{array}\right\} \tag{1.2.25}$$

$$\left.\begin{array}{ll} \text{bar S:} & \Delta\sigma = E(u/L) \\ \text{bar C:} & \Delta\sigma = E(\Delta u/L - \alpha T_Y) \end{array}\right\}. \tag{1.2.26}$$

Δu is determined from the equilibrium condition that the sum of forces induced in all bars should vanish. The result is expressed by the introduction of T_Y,

$$0 = A_0 E \Delta u/L + AE(\Delta u/L - \alpha T)$$
$$\Delta u = -\frac{A}{A + A_0}\alpha T_Y L = -\frac{A}{A_0}\frac{\sigma_Y L}{E}. \tag{1.2.27}$$

As this Δu coincides with u_Y, the displacement of the wall becomes zero, and the stresses in all bars vanish. Here, it should be noted that $\Delta\varepsilon^p = -\alpha(T_{max} - T_Y)$ remains in bar C, but this is cancelled by the thermal strain $\alpha(T_{max} - T_Y)$ of the bar, of which the temperature is $(T_{max} - T_Y)$. As a result, the total strain and stress in every bar become zero, and the stress point reaches C in Fig. 1.17.

② **Stage 2** (cooling to $T = 0°C$, i.e., from stress point C to D, Fig. 1.17).

Bar C is cooled down to $T = 0°C$ by $T_{max} - T_Y$. As the change of temperature is $\Delta T = (T_{max} - T_Y) \leq T_Y$, bar C is unloaded elastically, and the stress point reaches D in Fig. 1.18. The final strain and stress of each bar is calculated as follows.

The increments of total strain and stress of bar S are defined by

$$\Delta\varepsilon = \Delta u/L = \Delta\varepsilon^e$$
$$\Delta\sigma = E \cdot \Delta u/L.$$

As the total strain of bar C is composed of elastic and thermal strains, the increments of the total strain and stress of the bar yield

$$\Delta\varepsilon = \Delta u/L = \Delta\varepsilon^e - \alpha(T_{max} - T_Y) \tag{1.2.28}$$

$$\left.\begin{array}{l} \Delta\varepsilon^e = \Delta u/L + \alpha(T_{max} - T_Y) \\ \Delta\sigma = E\{\Delta u/L + \alpha(T_{max} - T_Y)\} \end{array}\right\}. \tag{1.2.29}$$

From the equilibrium condition, Δu is determined as

$$0 = A_0 \cdot \Delta u/L + A\{\Delta u/L + \alpha(T_{max} - T_Y)\}$$
$$\Delta u = -A/(A + A_0) \cdot \alpha(T_{max} - T_Y)L. \tag{1.2.30}$$

The stress changes induced in the three bars are produced by thermal strain $\alpha(T_{max} - T_Y)$ due to temperature change. Then, as the behaviors of the bars are elastic, their final stresses can be expressed by replacing αT in Eq. (1.2.19) by $\alpha(T_{max} - T_Y)$, as follows:

$$\left.\begin{array}{ll} \text{bar S:} & \sigma = -EA/(A + A_0) \cdot \alpha(T_{max} - T_Y) \\ \text{bar C:} & \sigma = E\{-A/(A + A_0) \cdot \alpha(T_{max} - T_Y) + \alpha(T_{max} - T_Y)\} \\ & = EA_0/(A + A_0) \cdot \alpha(T_{max} - T_Y) \end{array}\right\}. \quad (1.2.31)$$

At first glance, in Eq. (1.2.31) the source of residual stress seems to be the change of temperature at this step. However, the temperature returns to 0°C, and no thermal strain exists. Accordingly, the source of residual stress is the plastic strain, $\Delta\varepsilon^p = -\alpha(T_{max} - T_Y) \equiv \Delta\varepsilon_1^p$, which is produced at the heating stage. This plastic strain is inherent strain ε^* and integration of the plastic strain yields inherent displacement ΔL^*:

$$\varepsilon^* = \Delta\varepsilon_1^p = -\alpha(T_{max} - T_Y) \quad (1.2.32.A)$$

$$\Delta L^* = \varepsilon^* L = -\alpha(T_{max} - T_Y)L. \quad (1.2.32.B)$$

In this process, the maximum heating temperature is below $T_{max} = 2T_Y$, and the consequent plastic strain is $-\varepsilon_Y$ at most. At the subsequent cooling stage, bar C behaves elastically. If the maximum temperature is above $T_{max} = 2T_Y$, tensile plastic strain is induced at the cooling stage.

1.2.4.3 High-Temperature Heating
(Process ③, 0°C → $2T_Y \leq T_{max}$ → 0°C, Fig. 1.18)
Heating Stage (Heating from $T = 0°C$ to $2T_Y \leq T_{max}$, i.e., Stress Point O to P, Fig. 1.18)

In this case, the maximum temperature of bar C is above $2T_Y \leq T_{max}$, and the thermal strain reaches P' in the figure. As the bar is in the plastic range, the stress point stays at σ_Y and reaches P in the figure, and the additional thermal strain induced above T_Y,

$$\Delta\varepsilon^T = \alpha\,\Delta T = \alpha(T_{max} - T_Y),$$

is suppressed to turn into compressive plastic strain,

$$\Delta\varepsilon^p = -\Delta\varepsilon^T = -\alpha(T_{max} - T_Y) \equiv \Delta\varepsilon_1^p. \quad (1.2.33)$$

Cooling Stages

① **Stage 1** (cooling from $T_{max} > 2T_Y$ by T_Y, i.e., P to Q, Fig. 1.18).

At this cooling stage, the shrinkage strain of the bar is $-\alpha T_Y$, which reduces the stress by σ_Y and the stress in the bar vanishes. The stress point moves from P to Q.

② **Stage 2** (cooling further by T_Y to $T = T_{max} - 2T_Y$).

Shrinkage strain by this cooling is $-\alpha T_Y$, which is resisted by bar S through the wall and turns into tensile yield stress σ_Y. Then, the stress of bar C vanishes, and the stress point reaches R in Fig. 1.18.

③ **Stage 3** (cooling to $T = 0°C$).

At this stage, the shrinkage strain is $-\alpha(T_{max} - 2T_Y)$, and the strain reaches G′ in Fig. 1.18. As the stress is at a yielding point in tension, no further increase of stress occurs, and the shrinkage strain turns into tensile plastic strain. Stress point R shifts to G′, and the plastic strain increment is given by

$$\Delta \varepsilon^p = \alpha(T_{max} - 2T_Y) \equiv \Delta \varepsilon_2^p. \qquad (1.2.34)$$

The residual stresses in bars C and S are calculated as

$$\left. \begin{array}{ll} \text{bar C:} & \sigma = \sigma_Y \\ \text{bar S:} & \sigma = -(A/A_0)\,\sigma_Y \end{array} \right\}. \qquad (1.2.35)$$

In this process of high-temperature heating, plastic strains are produced at two stages: at the heating stage, $\Delta \varepsilon_1^p = -\alpha(T_{max} - T_Y)$, and at the cooling stage, $\Delta \varepsilon^p = \alpha(T_{max} - 2T_Y) \equiv \Delta \varepsilon_2^p$.

The former is compressive and the latter is tensile. Then, the sum of these two results in the residual plastic strain in bar C, that is,

$$\Delta \varepsilon^p = -\alpha(T_{max} - T_Y) + \alpha(T_{max} - 2T_Y) = -\alpha T_Y.$$

This plastic strain is inherent strain ε^*,

$$\varepsilon^* = \Delta \varepsilon_1^p + \Delta \varepsilon_2^p = -\alpha T_Y. \qquad (1.2.36.A)$$

The corresponding inherent deformation is given by

$$\Delta L^* = \varepsilon^* L = -\alpha T_Y L. \qquad (1.2.36.B)$$

As the displacement of the rigid body u coincides with that of bar S, it is provided by integration of the total strain of bar S as

$$u = -(\sigma_Y/E)(A/A_0)L. \qquad (1.2.37)$$

Here, the correlations between plastic strain, inherent strain, and inherent deformation, which are produced under the three different thermal processes, are summarized, and the result is completely identical to Table 1.4. Here, it should be noted that T_Y of this case should be $T_Y = \frac{A + A_0}{A_0}\frac{\sigma_Y}{E\alpha}$ for the base temperature 0°C, as shown in Eq. (1.2.21).

1.3 REPRODUCTION OF RESIDUAL STRESS BY INHERENT STRAIN AND INVERSE ANALYSIS FOR INHERENT STRAIN

1.3.1 Reproduction of Residual Stress by Inherent Strain

As realized through the analysis in the previous sections, the source of residual stress is the plastic strain resulting from the elastic-plastic behavior. Accordingly, if we know the appropriate distribution of inherent strain, which is the source of residual stress, we can reproduce the residual stress imposing the inherent strain to the stress-free model. This fact is represented in Table 1.4 as the relationship between plastic strain $\Delta\varepsilon^p$, inherent strain ε^*, inherent displacement ΔL^*, and residual stress σ. We can confirm this fact through the following calculations.

First, we apply a standard procedure of elastic analysis to calculate residual stress by imposing the inherent strain ε^* to bar C as loads. Practically, we can calculate it by replacing αT by ε^* in Eqs. (1.2.18) and (1.2.19). Consequently, the displacement of the rigid body u is obtained from Eq. (1.2.18) as

$$u = A/(A + A_0) \cdot \varepsilon^* L. \qquad (1.3.1)$$

And the stresses in bars C and S are from Eq. (1.2.19):

$$\left.\begin{array}{ll} \text{bar S:} & \sigma = E\dfrac{A}{A + A_0}\varepsilon^* \\[3mm] \text{bar C:} & \sigma = -E\dfrac{A_0}{A + A_0}\varepsilon^* \end{array}\right\}. \qquad (1.3.2)$$

By this example, we confirm that the proper inherent strain reproduces the proper residual stress. As illustrated in this example, we can apply this procedure to two- and three-dimensional objects to predict residual stresses at any points in the objects. In other words, if we construct a database accumulating information on inherent strains about appropriate types of welded joints, we can predict residual stresses of the joints by standard elastic analysis without performing complex thermal elastic-plastic analysis. This suggests the importance of analysis of inherent strain from existing residual stress distributions of welded joints. This method of analysis is presented in the next section and also in Chapter 2.

1.3.2 Inverse Analysis for Inherent Strain

Here, we apply the procedure explained in Section 1.3.1 to perform a numerical experiment on the model shown in Fig. 1.20 to obtain inherent strain.

As the first step, we apply a designed thermal history to the model manufactured for this experiment. Then, residual stresses are induced in the model. For measurement of residual stresses in the model, the following two methods may be used.

① Method of direct measurement: we measure the residual stresses of all three bars, as illustrated in Fig. 1.22(a).

② Method of inherent strain: we measure the residual stress of one of the three bars. Next, we estimate the inherent strain by inverse analysis, and we calculate the overall distribution of residual stresses in the three bars, as shown in Fig. 1.22(b).

By method ①, we can know the residual stresses at only the specified measuring points. We have to increase the number of measuring points where we need information. In contrast, by method ②, we can obtain overall information

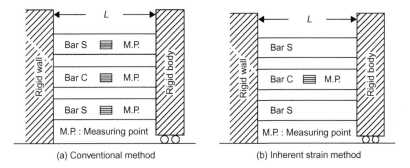

(a) Conventional method   (b) Inherent strain method

FIG. 1.22 Residual stress measurement by conventional method and inherent strain method.

about the residual stresses of the model. Here, we try to perform the inverse analysis on the model to obtain the inherent strains.

In this experiment, bar C is subjected to the thermal history, and presumably the inherent strain ε^* is induced in this bar. We measure the residual stress of any one of the bars, bar S for example, and observe the residual stress $_m\sigma$. This stress is expressed with the help of Eq. (1.3.2) in terms of ε^*, which is unknown at this stage:

$$_m\sigma = E\frac{A}{A+A_0}\varepsilon^*. \tag{1.3.3}$$

We solve this equation for ε^*:

$$\varepsilon^* = \frac{A+A_0}{A}\frac{_m\sigma}{E}. \tag{1.3.4}$$

We succeed in the inverse analysis. Substitution of this ε^* into Eq. (1.3.2) yields

$$\sigma = -\frac{A_0}{A}\,_m\sigma.$$

This is the exact value of the residual stress of bar C. We can start with the measurement of the stress of bar C instead of bar S and obtain the same residual stresses.

This example of the inherent strain method proves that the entire distribution of residual stresses can be obtained by measuring the residual stresses of a part of the object. In this method, inverse analysis is performed and the inherent strains are successfully found. In Chapter 2, the general procedure of this method is presented.

1.4 NUMERICAL EXAMPLES OF RESIDUAL STRESS, INHERENT STRAIN, AND INHERENT DISPLACEMENT

In Section 1.2.4, we analyzed the most general case of the three-bar model. The model is connected to the rigid wall on the left side and to the movable rigid body on the right side. Bar C of the model is subjected to the thermal process,

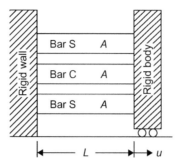

FIG. 1.23 Numerical example of elastically restrained bar.

$0°C \rightarrow T_{max} \rightarrow 0°C$, and its thermal elastic-plastic behavior is analyzed. According to the magnitude of T_{max}, the production process of residual stress is different. Consequently, the different magnitudes of residual stress and inherent displacement are produced.

To comprehend the complex behaviors of the model under the three different thermal processes, a numerical calculation is conducted. The processes are the same as in Section 1.2.4, which are low temperature, medium temperature and high temperature heating. For simplicity, the model assumes that the three bars have the same cross-sectional area, A.

Then, the new model becomes as shown in Fig. 1.23, and

$$A_0 = 2A.$$

1.4.1 Low-Temperature Heating Process ($T_{max} \leq T_Y$, Process ①, Fig. 1.21, also Refer to Fig. 1.16)

1.4.1.1 Heating Stage ($0°C \rightarrow T_{max}°C$)

Now, bar C is heated from $0°C$ to T_{max}. The displacement of the rigid body u and the stress of each bar are calculated from Eqs. (1.2.18) and (1.2.19) as

$$u = \frac{A}{A + A_0}\alpha TL = \frac{1}{3}\alpha TL \tag{1.4.1}$$

$$\left.\begin{array}{ll} \text{bar C:} & \sigma = -E\dfrac{A_0}{A + A_0}\alpha T = -\dfrac{2}{3}E\alpha T \\[3mm] \text{bar S:} & \sigma = E\dfrac{A}{A + A_0}\alpha T = \dfrac{1}{3}E\alpha T \end{array}\right\}. \tag{1.4.2}$$

The results are illustrated in Fig. 1.24(a). The temperature when the stress of bar C reaches the yield stress is the yield temperature T_Y that is given by Eq. (1.2.20),

$$T_Y = \frac{A + A_0}{A_0}\frac{\sigma_Y}{E\alpha} = \frac{3}{2}\frac{\sigma_Y}{E\alpha}. \tag{1.4.3}$$

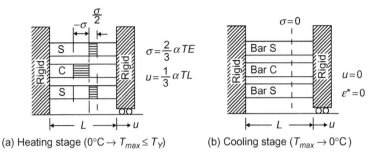

FIG. 1.24 Low-temperature heating process.

In this thermal process, T_{max} is kept lower than T_Y and no plastic strain is induced in bar C and bar S. Consequently, no plastic strain is produced in this heating process.

1.4.1.2 Cooling Stage ($T_{max}°C \to 0°C$)

When bar C is cooled down from T_{max} to $0°C$, no residual stress is found in bar C and bar S (Fig. 1.24(b)). This implies that there is neither inherent strain nor inherent displacement.

1.4.2 Medium-Temperature Heating (Process ②, Fig. 1.21, $T_Y \le T_{max} \le 2T_Y$, also Refer to Fig. 1.17)

1.4.2.1 Heating Stage ($0°C \to T_{max}°C$)

In the heating process to T_{max}, the stress of bar C reaches the compressive yield stress at T_Y, and the stress of each bar can be obtained by substituting Eq. (1.4.2) into Eq. (1.4.3):

$$\left. \begin{array}{ll} \text{bar C:} & \sigma = -\sigma_Y \\ \text{bar S:} & \sigma = \dfrac{1}{2}\sigma_Y \end{array} \right\}. \qquad (1.4.4)$$

These stresses are illustrated in Fig. 1.25(a). Above T_Y, the stress of bar C remains at the yielding stress, and bar C cannot sustain any more stress. Consequently, the movable rigid body will stay at $u_Y = \frac{1}{3}\alpha T_Y L$. The thermal strain increment above $T = T_Y$ is $\Delta \varepsilon^T = \alpha(T_{max} - T_Y)$, which becomes compressive plastic strain and reaches B′ in Fig. 1.17.

1.4.2.2 Cooling Stage ($T_{max}°C \to 0°C$)

① Stage 1 ($T_{max} \to (T_{max} - T_Y)$)

When bar C is cooled down from T_{max} to $T_{max} - T_Y$, the thermal strain (shrinkage) is $-\alpha T_Y$, and bar C and bar S are unloaded elastically. By this cooling, the movable rigid body displaces by Δu,

$$\Delta u = -\frac{A}{A + A_0}\alpha T_Y L = -\frac{1}{2}\varepsilon_Y L = -\frac{1}{2}\frac{\sigma_Y L}{E}. \qquad (1.4.5)$$

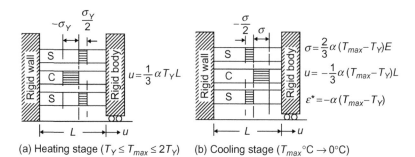

(a) Heating stage ($T_Y \leq T_{max} \leq 2T_Y$)   (b) Cooling stage ($T_{max}°C \to 0°C$)

FIG. 1.25   Medium-temperature heating process.

As this displacement coincides with u_Y, the total displacement of the rigid body becomes zero; in other words, the rigid body returns to the original position. In this state, the temperature of bar C is $T_{max} - T_Y$, and plastic strain $\Delta\varepsilon^p = -\alpha(T_{max} - T_Y)$ exists. The thermal strain and the plastic strain compensate each other. The result is that the total strain and the stresses in the bars vanish and the strain point reaches C in Fig. 1.17.

② **Stage 2 (($T_{max} - T_Y$) → 0°C)**

In this stage, bar C is cooled further from $T = T_{max} - T_Y$ to $T = 0°C$. As $T_{max} \leq 2T_Y$ in this stage, the temperature change ΔT is less than T_Y. Then, the bars are unloaded elastically and the strain point moves to D in Fig. 1.17. The displacement of the rigid body in this stage is

$$\Delta u = -\frac{1}{3}\alpha(T_{max} - T_Y)L. \tag{1.4.6}$$

The stress in each bar is shown as

$$\left.\begin{array}{ll} \text{bar C:} & \sigma = \dfrac{2}{3}E\alpha(T_{max} - T_Y) \\[2mm] \text{bar S:} & \sigma = -\dfrac{1}{3}E\alpha(T_{max} - T_Y) \end{array}\right\}. \tag{1.4.7}$$

As stated in the preceding sections, the source of residual stress is plastic strain induced in the heating stage $\Delta\varepsilon^p = -\alpha(T_{max} - T_Y) \equiv \Delta\varepsilon_1^p$. This is the inherent strain

$$\varepsilon^* = \Delta\varepsilon_1^p = -\alpha(T_{max} - T_Y). \tag{1.4.8}$$

Then, inherent displacement ΔL^* is given by

$$\Delta L^* = \varepsilon^*L = -\alpha(T_{max} - T_Y)L. \tag{1.4.9}$$

1.4.3 High-Temperature Heating (Process ③, Fig. 1.21, $T_{max} \geq 2T_Y$, also Refer to Fig. 1.18)

1.4.3.1 Heating Stage $(0°C \rightarrow T_{max})$

When the maximum temperature exceeds $2T_Y$, the thermal strain point moves to P′ in Fig. 1.18. As observed in the preceding example, the stress of bar C remains at σ_Y and that of bar S is

$$\left. \begin{array}{ll} \text{bar C:} & \sigma = -\sigma_Y \\ \text{bar S:} & \sigma = \dfrac{1}{2}\sigma_Y \end{array} \right\}. \tag{1.4.10}$$

These stresses are illustrated in Fig. 1.26(a). Thermal strain induced above T_Y becomes compressive plastic strain, that is,

$$\Delta \varepsilon^T = \alpha \Delta T = \alpha(T_{max} - T_Y) \equiv \Delta \varepsilon_1^p. \tag{1.4.11}$$

1.4.3.2 Cooling Stage

① **Stage 1** $(T_{max} \rightarrow (T_{max} - T_Y))$

As seen in the preceding examples, the shrinkage strain $-\alpha T_Y$ at this stage reduces the stress of each bar from σ_Y to zero, and the stress point reaches Q in Fig. 1.18.

② **Stage 2** $((T_{max} - T_Y) \rightarrow (T_{max} - 2T_Y))$

On this stage of cooling, the temperature decreases by T_Y and the shrinkage strain of bar C is $-\alpha T_Y$, which unloads bar C further from zero to the compressive yield stress. The strain point moves to R in Fig. 1.18.

③ **Stage 3** $((T_{max} - 2T_Y) \rightarrow 0°C)$

After further cooling from $(T_{max} - 2T_Y)$ to $0°C$, the strain point of bar C reaches G′ in Fig. 1.18. As the bar is yielded in tension, bar C cannot sustain any more stress and the stress point moves from R to G in Fig. 1.18. Consequently, the increment of the shrinkage strain turns into tensile plastic strain, that is,

$$\varepsilon^p = \alpha(T_{max} - 2T_Y) \equiv \Delta \varepsilon_2^p. \tag{1.4.12}$$

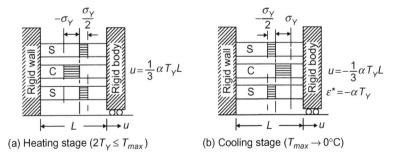

(a) Heating stage $(2T_Y \leq T_{max})$ (b) Cooling stage $(T_{max} \rightarrow 0°C)$

FIG. 1.26 High-temperature heating process.

The stress of each bar is calculated as follows and as illustrated in Fig. 1.26(b):

$$\left. \begin{array}{ll} \text{bar C:} & \sigma = -\sigma_Y \\ \text{bar S:} & \sigma = \dfrac{1}{2}\sigma_Y \end{array} \right\}. \tag{1.4.13}$$

In this thermal process of high-temperature heating, the following two kinds of plastic strain are produced: the compressive plastic strain $\Delta\varepsilon_1^p = -\alpha(T_{max} - T_Y)$ in the heating process, and the tensile plastic strain $\Delta\varepsilon_2^p = \alpha(T_{max} - 2T_Y)$ in the cooling stage. As a result, the summation of the two plastic strains induced on these processes remains in bar C:

$$\Delta\varepsilon^p = -\alpha(T_{max} - T_Y) + \alpha(T_{max} - 2T_Y) = -\alpha T_Y.$$

The inherent strain is written as

$$\varepsilon^* + \Delta\varepsilon_1^p = \Delta\varepsilon_2^p = -\alpha T_Y. \tag{1.4.14}$$

And the inherent displacement is written as

$$\Delta L^* = \varepsilon^* L = -\alpha T_Y L. \tag{1.4.15}$$

The displacement of the movable rigid body coincides with that of bar S, which is evaluated as

$$u = -\frac{\sigma_Y}{E}\frac{1}{2}L = -\frac{1}{3}\alpha T_Y L. \tag{1.4.16}$$

REFERENCES

[1] Kihara H, Masubuchi K. Welding deformation and residual stress. Tokyo: Sanpou Shupan; 1955 [in Japanese].

[2] Watanabe S, Satoh K. Welding mechanics and its application. Tokyo: Asakura Shoten; 1965 [in Japanese].

[3] Ueda Y, Yamakawa T. Analysis of thermal elastic-plastic stress and strain during welding by finite element method. Trans JWS 1971;2(2):90–100.

[4] Hibbitt HD, Marcal PV. A numerical thermo-mechanical model for the welding and subsequent loading of a fabricated structure. J Comput Struct 1973;3:1145–74.

[5] Satoh K, Ueda Y, Fujimoto T. Welding deformation and residual stress. Tokyo: Sanpo Shupan; 1979 [in Japanese].

[6] Ueda Y, Nakacho K, Kim Y, Murakawa H. Fundamentals of thermal-elastic-plastic-creep analysis and measurement of welding residual stress for numerical analysis (1st Report). J JWS 1986;55(6):336–48; (2nd Report). J JWS 1986;55(7):399–410; and (3rd report). J JWS 1986;55(8):458–65 [in Japanese].

[7] Japan Welding Society. Introduction to welding and joining. Tokyo: Sanpo Shupan; 1998 [in Japanese].

[8] Rykalin NN. Berechnung de Warmevorgange beim Schweissen. Berlin: VEB verlag Technik; 1957.

[9] Japan Association for High Pressure Technology. Standard manuals and guide for post-welding heat treatment. Tokyo: Nippon Kogyou Shinbun; 1994 [in Japanese].

Chapter 2

Introduction to Measurement and Prediction of Residual Stresses with the Help of Inherent Strains

When the safety of structures is discussed, information about welding distortion and residual stresses is important because it is the cause of many failures, including buckling, plastic collapse, fatigue, stress corrosion, etc.

As studied in Chapter 1, the residual stresses of a model can be obtained by measuring directly on the surfaces. Additionally, we also learned that the residual stresses at any point on the model can be estimated by standard elastic analysis if we have information about the inherent strains.

For measuring welding residual stresses, conventional methods include nondestructive methods, such as X-ray, γ-ray, the hole drilling method, the slicing method using strain gauges, etc. These methods provide limited information on residual stresses on the surface or in the limited depth from the surface. By these methods, we cannot measure three-dimensional welding residual stresses produced inside of an object.

Presently, we have to cut the object and expose the inside to measure three-dimensional residual stresses. For this purpose, two basic principles are discussed. One is the stress relaxation method, which is to repeatedly cut the object and measure the released stresses at each step of cutting. The sum of the released stresses is supposed to be the final residual stresses. This idea forms the basis of most conventional methods [1], but it is not accurate for all cases, since there are cases when we cannot measure all the components of released stresses.

Another method is the so-called inherent strain method [2–10], which allows us to accuately measure three-dimensional residual stress distribution. According to this method, we first measure the inherent strains (i.e., the source of the residual stresses) and then compute the entire distribution of welding residual stresses by elastic analysis using the measured inherent strains.

Welding Deformation and Residual Stress Prevention
© 2012 Elsevier Inc. All rights reserved.

In order to measure three-dimensional residual stresses produced inside of the object by any of the two methods stated above, we must cut and/or slice it. This processing induces change of stress distribution in the object. However, the inherent strain distribution would not change unless additional plastic strains were produced by the processing, since the main component of inherent strains is plastic strains.

The inherent method used is residual stress measurement. The first step is to cut the object into pieces in which the stress is in a plane stress state. The second step is to measure the residual stresses in the pieces, and then the inherent strains are calculated inversely from these measured stresses. The sum of the inherent strains in the pieces provides the entire distribution in the original object. The final step is to compute the entire distribution of residual stress in the object by elastic analysis, imposing the estimated inherent strains on the stress-free object. This method can be applied to measure not only two- but also three-dimensional residual stress distributions. Along these lines, some practical methods of measurement of three-dimensional residual stresses were developed, and details are described in references [2–10].

In this chapter, the basic theory of the inherent strain method and a measuring procedure for residual stresses is explained using a simple three-bar (one-dimensional) model [11].

As just stated, residual stresses can be reproduced by elastic calculation imposing the corresponding inherent strains on the stress-free body. This indicates inversely that we can predict welding residual stresses if we have information on the inherent strains of an intended type of joint, which can be obtained beforehand by experiments, theoretical analyses, etc. We can extend this fact to the prediction of welding deformation.

2.1 INHERENT STRAINS AND RESULTING STRESSES

Consider the model in Fig. 2.1, which is similar to the one in Fig. 1.9 in Chapter 1. The model is composed of the same three bars (bar 1, bar 2, and bar 3), each of which has cross-sectional area, A, length, L, and Young's modulus, E.

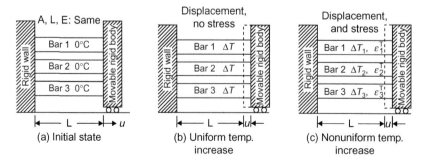

FIG. 2.1 Three-bar model for analysis.

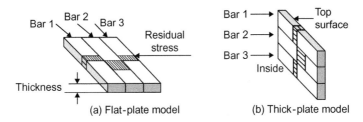

FIG. 2.2   Uniaxial stress models.

The three bars are fixed at one end and rigidly connected to a movable rigid body. This model may be regarded as the following two practical cases, as shown in Fig. 2.2(a) and (b), respectively:

1. An idealized plane model of a butt-welded joint, Fig. 2.2(a); bar 2 corresponds to the welded metal and bars 1 and 3 to the adjacent plain plates.
2. An idealized three-dimensional model in a welded joint of thick plates, Fig. 2.2(b); bar 1 is located at the upper surface, bar 2 in the center, and bar 3 on the lower surface.

In Chapter 1, the relation among thermal deformation and stress, and inherent strain under the specified thermal processes was described.

Here, Fig. 2.1(a) is the initial state, and we heat the three bars in different ways:

1. As shown in Fig. 2.1(b), if we heat up the three bars to the same temperature thermal elongations of the three bars are the same and no interaction among them can not be recognized through the movable rigid body, like free bars as Section 1.2.2. Consequently no thermal stresses are produced in the bars. Although the thermal strain is a kind of inherent strain, in this case does not cause the stresses but causes the deformation.
2. If we only warm up bar 2, the bar may elongate due to thermal strain, but the other bars resist the elongation of bar 2. This causes thermal stress in bar 2, and reaction stresses in bars 1, and 3. This fact suggests that the difference between the thermal strains in the three bars becomes the source of thermal stresses. In this case, this main source is regarded as inherent strain.

What we learn from the above examples is that depending on the constraint condition against thermal strain, inherent strain (thermal strain) becomes effective to cause thermal stress or non-effective.

If we know the magnitude and distribution of the inherent strains, we can calculate thermal stress and deformation by a standard procedure of elastic analysis, imposing the inherent stains as external loads.

In the following, the basic relations necessary for the analysis of this example are formulated.

To treat the problem more generally, different magnitudes of temperature increase ΔT_1, ΔT_2, and ΔT_3 are specified for bar 1, bar 2, and bar 3. The thermal strains in bar 1, bar 2, and bar 3 are ε_1^T, ε_2^T, and ε_3^T, which may be obtained as a product of the temperature increase and thermal expansion coefficient α. The thermal elongations of bar 1, bar 2, and bar 3 are denoted by u_1^T, u_2^T, and u_3^T, which are given by integration of thermal strain along the length, as follows:

$$\{\varepsilon^T\} = \alpha\{\Delta T\}, \tag{2.1.1}$$

$$\{u^T\} = \{\varepsilon^T\}L = \alpha L\{\Delta T\}, \tag{2.1.2}$$

where

$$\begin{aligned}
\{\varepsilon^T\} &= \begin{bmatrix} \varepsilon_1^T & \varepsilon_2^T & \varepsilon_3^T \end{bmatrix}^T \\
\{u^T\} &= \begin{bmatrix} u_1^T & u_2^T & u_3^T \end{bmatrix}^T \\
\{\Delta T\} &= \begin{bmatrix} \Delta T_1 & \Delta T_2 & \Delta T_3 \end{bmatrix}^T \\
&[\ ]^T : \text{transposed matrix of } [\ ].
\end{aligned} \tag{2.1.3}$$

As the three bars of the model are connected to the movable rigid body, the free thermal elongation of each bar is resisted by the interaction of the three bars, and the resisting forces produce stress in each bar. In this state, the movable rigid body would move by u, which is the same displacement of the three bars, so that the resisting forces in the three bars should be in equilibrium. Then, as long as the temperature increase is unchanged, these thermal strains are the source of stresses and can be regarded as inherent strains. Here, the thermal strains $\{\varepsilon^T\}$ are replaced by $\{\varepsilon^*\}$, that is,

$$\{\varepsilon^*\} = \{\varepsilon^T\}, \tag{2.1.4}$$

where $\{\varepsilon^*\} = \begin{bmatrix} \varepsilon_1^* & \varepsilon_2^* & \varepsilon_3^* \end{bmatrix}^T$.

Now we will calculate the thermal stresses in the bars produced by the inherent strain $\{\varepsilon^*\}$ by the standard procedure of theory of elasticity.

Here, we rewrite the equations derived in Section 1.2.

2.1.1 Displacement–Strain Relation (Compatibility)

As the displacement of the bars is u, the total strain $\{\varepsilon\}$ is defined as

$$\{\varepsilon\} = \{u\}/L, \tag{2.1.5}$$

where

$$\begin{aligned}
\{\varepsilon\} &= \begin{bmatrix} \varepsilon & \varepsilon & \varepsilon \end{bmatrix}^T \\
\{u\} &= \begin{bmatrix} u & u & u \end{bmatrix}^T.
\end{aligned}$$

The total strain $\{\varepsilon\}$ is composed of elastic strain $\{\varepsilon^e\}$ and inherent strain $\{\varepsilon^*\}$ as

$$\{\varepsilon\} = \{\varepsilon^e\} + \{\varepsilon^*\}, \tag{2.1.6}$$

where $\{\varepsilon^e\} = \begin{bmatrix} \varepsilon_1^e & \varepsilon_2^e & \varepsilon_3^e \end{bmatrix}^T$.

Then, the elastic strain is given as

$$\{\varepsilon^e\} = \{\varepsilon\} - \{\varepsilon^*\}. \tag{2.1.7}$$

2.1.2 Stress–Strain Relation (Constitution Equation)

Stress $\{\sigma\}$ of the bar is expressed in terms of the elasticity matrix $[D]$ by the equation

$$\{\sigma\} = [D]\{\varepsilon^e\}, \tag{2.1.8}$$

where

$$\{\sigma\} = [\sigma_1 \quad \sigma_2 \quad \sigma_3]^T$$

$$[D] = \begin{bmatrix} E & 0 & 0 \\ 0 & E & 0 \\ 0 & 0 & E \end{bmatrix} = E \begin{bmatrix} 1 & 0 & 0 \\ 0 & 1 & 0 \\ 0 & 0 & 1 \end{bmatrix} = E[I] \tag{2.1.9}$$

$[I]$: unit matrix.

2.1.3 Equilibrium Condition (Equilibrium Equation)

The equilibrium equation can be derived from the condition that the sum of the forces of the three bars should vanish. As the cross-sectional area of each bar is the same as A, the condition is expressed as

$$\begin{aligned} 0 &= \{A\}^T\{\sigma\} = \{A\}^T[D]\{\varepsilon^e\} = \{A\}^T[D](\{\varepsilon\} - \{\varepsilon^*\}) \\ &= AE[1 \quad 1 \quad 1](\{\varepsilon\} - \{\varepsilon^*\}) \\ &= AE(3\varepsilon - (\varepsilon_1^* + \varepsilon_2^* + \varepsilon_3^*)), \end{aligned} \tag{2.1.10}$$

where $\{A\} = [A \quad A \quad A]^T$.

Substituting Eqs. (2.1.8), (2.1.7), and (2.1.5) into Eq. (2.1.10), we obtain total strain ε and displacement u as

$$\varepsilon = \frac{1}{3}(\varepsilon_1^* + \varepsilon_2^* + \varepsilon_3^*) \tag{2.1.11}$$

$$u = \varepsilon L = \frac{L}{3}(\varepsilon_1^* + \varepsilon_2^* + \varepsilon_3^*). \tag{2.1.12}$$

With these expressions, the resulting elastic strain $\{\varepsilon^e\}$ due to inherent strain $\{\varepsilon^*\}$ can be expressed by inserting Eq. (2.1.11) into Eq. (2.1.7) as

$$\{\varepsilon^e\} = \{\varepsilon\} - \{\varepsilon^*\}. \tag{2.1.13}$$

Equation (2.1.13) may be expressed in terms of $\{\varepsilon^*\}$ as

$$\{\varepsilon^e\} = [H^*]\{\varepsilon^*\} \quad \text{: elastic response equation,} \tag{2.1.14}$$

where

$$[H^*] = \frac{1}{3}\begin{bmatrix} -2 & 1 & 1 \\ 1 & -2 & 1 \\ 1 & 1 & -2 \end{bmatrix} \quad \text{: elastic response matrix.} \tag{2.1.15}$$

Finally, we obtain the stress $\{\sigma\}$ from Eq. (2.1.8) using Eq. (2.1.14) as

$$\{\sigma\} = [D]\{\varepsilon^e\} = [D][H^*]\{\varepsilon^*\}. \tag{2.1.16}$$

The characteristics of the elastic response matrix $[H^*]$ is described later in Section 2.6.

2.2 MEASURED STRAINS IN EXPERIMENTS AND INHERENT STRAINS

In the previous section, it was shown that the resulting elastic stress $\{\sigma\}$ due to inherent strain $\{\varepsilon^*\}$ can be obtained by a standard procedure of elastic analysis. Now we consider inversely how to obtain inherent strains by experiments. To this end, we should find a method to estimate the inherent strains from the measurement of surface elastic strains of specimens, which is the inverse procedure of the standard method described in the previous section. Here, we develop a procedure to determine the inherent strains by the experiment.

The following two methods are used to observe elastic strains (stresses) on the surface of a specimen, as illustrated in Fig. 2.3:

- Direct measurement of surface elastic strains by any method, Fig. 2.3(a). Measured strain is denoted by $\{_m\varepsilon^e\}$.
- Stress relaxation method, Fig. 2.3(b). First, strain gauges are put on the surface of the specimen, stresses are released by cutting the specimen near the strain gauges, and the released elastic strains are measured. Observed released strains with the opposite sign correspond to the surface elastic strains. This is denoted by $\{_m\varepsilon^e\}$ which is defined in the same way as in the previous case.

In the following, we perform measurements on the model shown in Fig. 2.1, assuming that the measured strains contain no experimental error. The case where experimental error is inevitable is treated in Section 2.5.

The resulting residual stress $\{\sigma\}$ due to inherent strain $\{\varepsilon^*\}$ is expressed by Eq. (2.1.16), and this may be rewritten in terms of $\{\varepsilon^e\}$, as shown by Eq. (2.1.14). As no experimental error is assumed, the measured strain $\{_m\varepsilon^e\}$

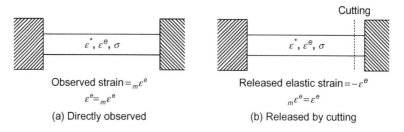

(a) Directly observed — Observed strain $=_m\varepsilon^e$, $\varepsilon^e =_m\varepsilon^e$

(b) Released by cutting — Cutting, Released elastic strain $=-\varepsilon^e$, $_m\varepsilon^e = \varepsilon^e$

FIG. 2.3   Conventional methods of elastic strain measurement by experiment.

should be equivalent to the residual elastic strain $\{\varepsilon^e\}$ in the specimen. Then, Eq. (2.1.14) becomes Eq. (2.2.1), which is called the observation equation:

$$\{_m\varepsilon^e\} = [H^*]\{\varepsilon^*\} \quad : \text{the observation equation.} \tag{2.2.1}$$

The equation contains three unknown quantities in matrix $\{\varepsilon^*\}$ and three measured strains in matrix $\{_m\varepsilon^e\}$. From this, $[H^*]$ is a square matrix, and the equation composes a set of simultaneous equations against $\{\varepsilon^*\}$.

Multiplication of the inverse matrix of $[H^*]$, that is, $[H^*]^{-1}$, to both sides of the equation yields

$$[H^*]^{-1}\{_m\varepsilon^e\} = [H^*]^{-1}[H^*]\{\varepsilon^*\} = [I]\{\varepsilon^*\}.$$

And then,

$$\{\varepsilon^*\} = [H^*]^{-1}\{_m\varepsilon^e\}. \tag{2.2.2}$$

To obtain the solution of the previous equation, we must calculate $[H^*]^{-1}$. A question arises about whether $\det[H^*]$ exists or not. We perform actual calculation using Eq. (2.1.15) and find the result as

$$\det[H^*] = 0. \tag{2.2.3}$$

This implies that we cannot obtain the solution. In the next section, we will clarify the reason and develop a method for estimation of $\{\varepsilon^*\}$.

2.3 EFFECTIVE AND NONEFFECTIVE INHERENT STRAINS

In Section 2.1 we loaded inherent strain $\{\varepsilon^*\}$ on the three bars. In this section, we impose an additional inherent strain $\{\Delta\varepsilon^*\}$ of the same magnitude to all bars, as shown in Fig. 2.4.

By the addition of $\{\Delta\varepsilon^*\}$, the total inherent stain $\{_t\varepsilon^*\}$ becomes Eq. (2.3.1), and additional displacement $\{\Delta u\}$ of the movable rigid body is computed by Eq. (2.3.2):

$$\{_t\varepsilon^*\} = \{\varepsilon^*\} + \{\Delta\varepsilon^*\} \tag{2.3.1}$$

$$\{\varepsilon\} = \{\varepsilon^e\} + \{_t\varepsilon^*\} = \{\varepsilon^e\} + \{\varepsilon^*\} + \{\Delta\varepsilon^*\}$$
$$\{u\} + \{\Delta u\} = [\{\varepsilon^e\} + \{\varepsilon^*\} + \{\Delta\varepsilon^*\}]L, \tag{2.3.2}$$
$$\{\Delta u\} = \{\Delta\varepsilon^*\}L$$

where

$$\{\Delta\varepsilon^*\} = [\Delta\varepsilon^* \quad \Delta\varepsilon^* \quad \Delta\varepsilon^*]^T$$
$$\{_t\varepsilon^*\} = [_t\varepsilon_1^* \quad _t\varepsilon_2^* \quad _t\varepsilon_3^*]^T$$

Here, we substitute the total inherent strain $\{_t\varepsilon^*\}$ into Eq. (2.1.15) to calculate stresses in the bars due to the additional inherent strain $\{\Delta\varepsilon^*\}$.

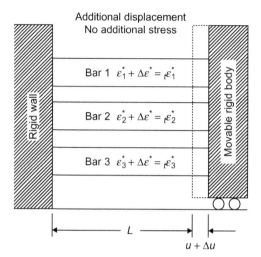

FIG. 2.4   Displacement of the rigid body by additional inherent strain.

The resulting stresses are the same as those induced by $\{\varepsilon^*\}$. From this fact, it turns out that $\{\Delta\varepsilon^*\}$ does not produce additional stress, but produces deformation. This case corresponds to the case shown in Fig. 2.1(b), where the same temperature increase is given to the three bars. Here, we recognize that there are two kinds of components in inherent strain: effective and noneffective. The effective components are produced only by differences among the inherent strains in the three bars, and the base value for the comparison is arbitrary, so we can take $\varepsilon_3^* = 0$.

Now we intend to determine effective inherent strains from the measured elastic strains of the specimen. As the procedure should be the inverse of the previous standard method, the noneffective component $\{\Delta\varepsilon^*\}$ included in the total inherent strains would not be determined from the measured strain, since the noneffective component has no relation with the stresses (the measured strains) in the previous case. This is the main reason we cannot solve Eq. (2.2.2).

To conduct the inverse analysis, we must eliminate the noneffective component from the measured elastic strains. One way is to specify the value of one component of $\{\varepsilon^*\}$ as a base value, and the relative difference of $\{\varepsilon^*\}$ of other bars can be calculated.

2.4 DETERMINATION OF EFFECTIVE INHERENT STRAINS FROM MEASURED RESIDUAL STRESSES

Here, we are going to calculate the effective inherent strains and estimate the resulting stresses. As the first step, we set the inherent strain of bar 3, ε_3^*, as the base value and regard the relative difference of the inherent strains of the other bars as effective ones. They can be shown as

$$\{\varepsilon^*\} - \{\varepsilon_3^*\} = \begin{Bmatrix} \varepsilon_1^* \\ \varepsilon_2^* \\ \varepsilon_3^* \end{Bmatrix} - \begin{Bmatrix} \varepsilon_3^* \\ \varepsilon_3^* \\ \varepsilon_3^* \end{Bmatrix} = \begin{Bmatrix} \varepsilon_1^* - \varepsilon_3^* \\ \varepsilon_2^* - \varepsilon_3^* \\ 0 \end{Bmatrix}, \tag{2.4.1}$$

where $\{\varepsilon_3^*\} = [\varepsilon_3^* \quad \varepsilon_3^* \quad \varepsilon_3^*]^T$.

For simpler manipulation in the following, the effective inherent strain $\{\varepsilon^*\} - \{\varepsilon_3^*\}$ is replaced by $\{\varepsilon^*\}$. Then,

$$\{\varepsilon^*\} = [\varepsilon_1^* \quad \varepsilon_2^* \quad 0]^T. \tag{2.4.2}$$

This indicates that effective inherent strains exist only in bars 1 and 2. This is illustrated in Fig. 2.5.

Next, substituting Eq. (2.4.2) into the right side of Eq. (2.1.14) we obtain the elastic response equation (2.4.3) with respect to the two effective inherent strains of $\{\varepsilon^*\}$ as

$$\{\varepsilon^e\} = \{\varepsilon\} - \{\varepsilon^*\} = [H^*]\{\varepsilon^*\}, \tag{2.4.3}$$

where

$$[H^*] = \frac{1}{3} \begin{bmatrix} -2 & 1 \\ 1 & -2 \\ 1 & 1 \end{bmatrix}. \tag{2.4.4}$$

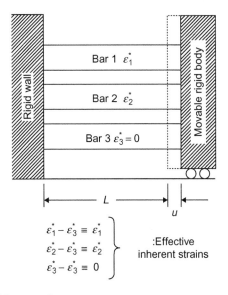

FIG. 2.5   Effective inherent strains.

Finally, we obtain the residual stresses $\{\sigma\}$ due to the effective inherent strains as

$$\{\sigma\} = [D]\{\varepsilon^e\} = [D][H^*]\{\varepsilon^*\}. \tag{2.4.5}$$

In Section 2.2, it was demonstrated that inherent strains cannot be determined from the measured elastic strains induced by the total inherent strains by an inverse method of standard analysis, and the reason was explained in Section 2.3.

In this section, then, it is shown that effective inherent strains can be determined from the residual stress indicated by Eq. (2.4.5), which are produced by the effective inherent strains if we use the proper inverse procedure.

$\{\varepsilon^e\}$ in Eq. (2.4.3) is the elastic strains developed in the bars. Here, we assume these strains are the measured strains and replace by $\{_m\varepsilon^e\}$. Then, we obtain

$$\begin{Bmatrix} _m\varepsilon_1^e \\ _m\varepsilon_2^e \\ _m\varepsilon_3^e \end{Bmatrix} = \frac{1}{3} \begin{bmatrix} -2 & 1 \\ 1 & -2 \\ 1 & 1 \end{bmatrix} \begin{Bmatrix} \varepsilon_1^* \\ \varepsilon_2^* \\ 0 \end{Bmatrix}, \tag{2.4.6}$$

and in matrix form,

$$\{_m\varepsilon^e\} = [H^*]\{\varepsilon^*\}, \tag{2.4.7}$$

where $\{_m\varepsilon^e\} = [\,_m\varepsilon_1^e \quad _m\varepsilon_2^e \quad _m\varepsilon_3^e\,]^T$.

This provides three equations for the two unknowns ε_1^* and ε_2^*. The elastic response matrix is not square and we cannot calculate $[H^*]^{-1}$. A general method for the solution is explained in the following section.

For this specific problem, we can obtain the two unknowns from any two equations out of three as a set of simultaneous equations. For example, we adopt the first two equations of Eq. (2.4.6), which include the measured strains of bars 1 and 2, and the two unknowns, and we solve them for the effective inherent strains:

$$\begin{Bmatrix} _m\varepsilon_1^e \\ _m\varepsilon_2^e \end{Bmatrix} = \frac{1}{3} \begin{bmatrix} -2 & 1 \\ 1 & -2 \end{bmatrix} \begin{Bmatrix} \varepsilon_1^* \\ \varepsilon_2^* \end{Bmatrix}. \tag{2.4.8}$$

Then, the solution is

$$\begin{Bmatrix} \varepsilon_1^* \\ \varepsilon_2^* \end{Bmatrix} = \begin{bmatrix} -2 & -1 \\ -1 & -2 \end{bmatrix} \begin{Bmatrix} _m\varepsilon_1^e \\ _m\varepsilon_2^e \end{Bmatrix}. \tag{2.4.9}$$

This example proves that we can obtain the effective inherent strains by the inverse method of the standard procedure. It is obvious that the effective inherent strains $\{\varepsilon^*\}$ produces the same residual stresses shown by Eq. (2.4.5). We can obtain the same results starting with a combination of any two equations of Eq. (2.4.6).

If the total inherent strains are required, we must measure the displacement of the movable rigid body u and determine the magnitude of the inherent strain ε_3^* specified as the base value with the help of Eqs. (2.1.5) and (2.3.1).

As illustrated earlier, we can obtain the exact values of the effective inherent strains from the measured elastic strains when there are no experimental errors. In general cases, various kinds of errors are included in the measurement. When errors are inevitable in measured elastic strains, we may get different inherent strains calculated from an arbitrary combination of two equations out of the three. The solution to this problem is explained in the next section.

2.5 MOST PROBABLE VALUE OF EFFECTIVE INHERENT STRAIN AND ACCURACY OF THE MEASUREMENT OF RESIDUAL STRESS

In this section, a procedure to determine the effective inherent strains is explained for the case where measured elastic strains contain some errors in experiments.

First, denoting the observation error by $\{\Delta_m \varepsilon^e\}$ with respect to the true elastic strain $\{\varepsilon^e\}$, the measured elastic strain $\{_m\varepsilon^e\}$ is expressed by the sum of $\{\varepsilon^e\}$ and $\{\Delta_m \varepsilon^e\}$ as

$$\{_m\varepsilon^e\} = \{\varepsilon^e\} + \{\Delta\varepsilon^e\}, \tag{2.5.1}$$

where

$$\{\Delta\varepsilon^e\} = [\Delta\varepsilon_1^e \quad \Delta\varepsilon_2^e \quad \Delta\varepsilon_3^e]^T.$$

Substituting $\{\varepsilon^e\}$ of Eq. (2.4.3) into the above, $\{_m\varepsilon^e\}$ is rewritten as

$$\{_m\varepsilon^e\} = [H^*]\{\varepsilon^*\} + \{\Delta\varepsilon^e\}. \tag{2.5.2}$$

Instead of the true inherent strain, the most probable value of the effective inherent strain $\{\hat{\varepsilon}^*\}$ is expressed as

$$\{_m\varepsilon^e\} - [H^*]\{\hat{\varepsilon}^*\} = \{v\}, \tag{2.5.3}$$

where

$$\{\hat{\varepsilon}^*\} = [\hat{\varepsilon}_1^* \quad \hat{\varepsilon}_2^* \quad \hat{\varepsilon}_3^*]^T$$
$$\{v\} = [v_1 \quad v_2 \quad v_3]^T : \text{residuals}.$$

The most probable value of the effective inherent strains $\{\hat{\varepsilon}^*\}$ may be found by applying the method of least squares. The condition may be rephrased as the sum of squares of the residuals, S, should be minimized. S is shown as

$$S = \{v\}^T\{v\}. \tag{2.5.4}$$

Then, the condition is

$$\left\{\frac{\partial S}{\partial \hat{\varepsilon}^*}\right\} = 0. \tag{2.5.5}$$

From Eq. (2.5.5), the most probable value is given as

$$\{\hat{\varepsilon}^*\} = ([H^*]^T[H^*])^{-1}[H^*]^T\{{}_m\varepsilon^e\}, \tag{2.5.6}$$

where

$$([H^*]^T[H^*])^{-1}[H^*]^T = \begin{bmatrix} -1 & 0 & 1 \\ 0 & -1 & 1 \end{bmatrix}.$$

Then, further manipulation of Eq. (2.5.6) leads to the expression

$$\begin{Bmatrix} \hat{\varepsilon}_1^* \\ \hat{\varepsilon}_2^* \end{Bmatrix} = \begin{bmatrix} -1 & 0 & 1 \\ 0 & -1 & 1 \end{bmatrix} \begin{Bmatrix} {}_m\varepsilon_1^e \\ {}_m\varepsilon_2^e \\ {}_m\varepsilon_3^e \end{Bmatrix} = \begin{Bmatrix} {}_m\varepsilon_3^e - {}_m\varepsilon_1^e \\ {}_m\varepsilon_3^e - {}_m\varepsilon_1^e \end{Bmatrix}. \tag{2.5.7}$$

As a result, the most probable values of the elastic strain and residual stress are derived as

$$\{\hat{\varepsilon}^e\} = [H^*]\{\hat{\varepsilon}^*\}$$
$$\{\hat{\sigma}\} = [D]\{\hat{\varepsilon}^e\} = [D][H^*]\{\hat{\varepsilon}^*\}. \tag{2.5.8}$$

In the preceding expressions, $[H^*]$ is given by Eq. (2.4.4).

Equations (2.5.6) and (2.5.8) are general expressions to be applied to two- and three-dimensional problems. However, attention should be paid to select appropriate elastic response matrices for respective problems.

In the case where measured strain $\{{}_m\varepsilon^e\}$ does not contain any errors, that is, $\{\Delta_m\varepsilon^e\} = \{0\}$, $\{{}_m\varepsilon^e\}$ in Eq. (2.5.1) is equal to elastic strain $\{\varepsilon^e\}$ and naturally the most probable values should be the true ones.

The accuracy of $\{\hat{\varepsilon}^e\}$ or $\{\hat{\sigma}\}$ may be evaluated in the following ways. They may be compared with directly measured elastic strain $\{{}_m\varepsilon^e\}$ or $\{{}_m\sigma\}$, which are observed on bars 1, 2, and 3. Another method is from the unbiased estimate of measurement variance $\hat{s}^2$, which is

$$\hat{s}^2 = S/(m-q). \tag{2.5.9}$$

Here, m is the number of measured strains and q is the number of unknown inherent strains.

2.6 DERIVATION OF ELASTIC RESPONSE MATRIX

When we estimate the inherent strains from measured strains, we need to calculate the elastic response matrix $[H^*]$. Although $[H^*]$ in Eq. (2.4.5) is derived analytically, it may not be applicable to general complex cases. To overcome the difficulty, a method for how to derive it by numerical analysis, for example, by the finite element method, is discussed.

Again, Eq. (2.4.3) is

$$\{\varepsilon^e\} = [H^*]\{\varepsilon^*\}. \tag{2.6.1}$$

This equation indicates that we can calculate elastic strain $\{\varepsilon^e\}$ at any point induced by the inherent strain $\{\varepsilon^*\}$. If the inherent strain is assumed to be $\{\varepsilon^*\} = [1 \quad 0]^T$, the resulting strains are

$$\begin{Bmatrix} \varepsilon_1^e \\ \varepsilon_2^e \\ \varepsilon_3^e \end{Bmatrix} = \begin{Bmatrix} -2/3 \\ 1/3 \\ 1/3 \end{Bmatrix}. \tag{2.6.2}$$

This is exactly the same as the first column of $[H^*]$ of Eq. (2.4.4). In the same manner, the result provides the second column for $\{\varepsilon^*\} = [0 \quad 1]^T$. In this way, we develop $[H^*]$ for this case. This procedure can be applied to general cases.

Here, for a general case where q number of inherent strains is given and elastic strains are measured at m number of points, Eq. (2.4.3) is expressed as

$$\begin{Bmatrix} \varepsilon_1^e \\ \varepsilon_2^e \\ \cdot \\ \cdot \\ \varepsilon_m^e \end{Bmatrix} = \begin{bmatrix} h_{11} & h_{12} & \cdot & h_{1q} \\ h_{21} & & \cdot & \cdot \\ \cdot & \cdot & \cdot & \cdot \\ \cdot & \cdot & \cdot & \cdot \\ h_{m1} & \cdot & \cdot & h_{mq} \end{bmatrix} \begin{Bmatrix} \varepsilon_1^* \\ \varepsilon_2^* \\ \cdot \\ \cdot \\ \varepsilon_q^* \end{Bmatrix}. \tag{2.6.3}$$

If we impose $\{\varepsilon^*\} = [1 \quad 0 \quad 0 \quad \cdots \quad 0]^T$, elastic strain $\{\varepsilon^e\}$ caused at the measuring points is given as

$$\begin{Bmatrix} \varepsilon_1^e \\ \varepsilon_2^e \\ \cdot \\ \cdot \\ \varepsilon_m^e \end{Bmatrix} = \begin{bmatrix} h_{11} \\ h_{21} \\ \cdot \\ \cdot \\ h_{m1} \end{bmatrix}. \tag{2.6.4}$$

This corresponds to the first column of $[H^*]$. We use the same procedure, giving the following conditions

$$\{\varepsilon^*\} = [1 \quad 0 \quad 0 \quad \cdots \quad 0]^T$$
$$\{\varepsilon^*\} = [0 \quad 1 \quad 0 \quad \cdots \quad 0]^T$$
$$\{\varepsilon^*\} = [0 \quad 0 \quad 1 \quad \cdots \quad 0]^T$$
$$\{\varepsilon^*\} = [0 \quad 0 \quad 0 \quad \cdots \quad 1]^T.$$

We can calculate all components of $[H^*]$.

When we deal with practical problems, the m number of measured strains should be equal to or larger than the q number of unknown inherent strains. Then, we must classify it into three cases comparing the numbers q against m:

1. In the case of $q = m$, $[H^*]$ is square, and the unknown inherent strains may be obtained as the solution of the simultaneous equations.

2. In the case of $q \leq m$, we can obtain most reliable values of the unknown inherent strains by reliability analysis.

3. In the case of $q > m$, we should increase measured strains so as to satisfy the above condition in case 1 or 2.

2.7 MEASURING METHODS AND PROCEDURES OF RESIDUAL STRESSES IN TWO- AND THREE-DIMENSIONAL MODELS

In the preceding sections, two basic analysis methods were represented: the standard method to calculate residual stresses induced by the inherent strains, and the inverse method to identify the inherent strains from the measured residual stresses (strains). Based on these methods and procedures, we will measure residual stresses in the two idealized models illustrated in Section 2.1: the two-dimensional model and the three-dimensional model. Both models are composed of three bars, and the effective inherent strains are assumed to exist in bars 1, and 2, as explained in Section 2.4. Then, more than two measured strains are necessary to determine the two effective inherent strains.

2.7.1 Measurement of Two-Dimensional Residual Stresses Induced in the Butt-Welded Joint of a Plate

We can measure residual stresses of the butt-welded joints of a plate using the measuring procedure explained in Sections 2.1 to 2.5.

As we work with the model of Fig. 2.2(a), which is an idealized butt-welded joint of a flat plate, we again idealize it as illustrated in Fig. 2.6. First, we put strain gauges directly on the surfaces. Next, we cut the three bars simultaneously to release the constrained stresses in the bars and measure the released strains. The measured released strains of the reverse sign are denoted by $_m\varepsilon_1^e$, $_m\varepsilon_2^e$, and $_m\varepsilon_3^e$.

When the measured strains are free from any experimental error, the procedure and corresponding results are completely the same as those in Section 2.4. Here, we rewrite the results. First, we obtain Eq. (2.4.6) by substitution of these measured strains into the right side of Eq. (2.4.3), that is,

$$\begin{Bmatrix} _m\varepsilon_1^e \\ _m\varepsilon_2^e \\ _m\varepsilon_3^e \end{Bmatrix} = \frac{1}{3} \begin{bmatrix} -2 & 1 \\ 1 & -2 \\ 1 & 1 \end{bmatrix} \begin{Bmatrix} \varepsilon_1^* \\ \varepsilon_2^* \\ 0 \end{Bmatrix}. \tag{2.4.6}$$

This set of equations consists of three equations with two unknown quantities, ε_1^* and ε_2^*. Then, we can solve these equations for ε_1^* and ε_2^* using any two equations out of three and obtain the same inherent strains. If we take the first two equations, the effective inherent strains are obtained as shown below, that is Eq. (2.4.9);

$$\begin{Bmatrix} \varepsilon_1^* \\ \varepsilon_2^* \end{Bmatrix} = \begin{bmatrix} -2 & -1 \\ -1 & -2 \end{bmatrix} \begin{Bmatrix} _m\varepsilon_1^e \\ _m\varepsilon_2^e \end{Bmatrix}. \tag{2.4.9}$$

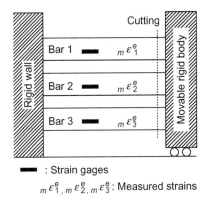

Cutting

Rigid wall

Bar 1 ▬ $_m\varepsilon_1^e$

Bar 2 ▬ $_m\varepsilon_2^e$

Bar 3 ▬ $_m\varepsilon_3^e$

Movable rigid body

▬ : Strain gages

$_m\varepsilon_1^e, _m\varepsilon_2^e, _m\varepsilon_3^e$: Measured strains

FIG. 2.6 Measuring procedure of two-dimensional residual stresses (flat-plate model).

Next, when the measured strains contain errors, we will obtain the most probable values of effective inherent strains. We have three measured strains—$_m\varepsilon_1^e$, $_m\varepsilon_2^e$, and $_m\varepsilon_3^e$—against the two unknowns. Following the detailed procedure discussed in Section 2.5, we find the most probable values of $\{\hat{\varepsilon}^*\}$ by Eq. (2.5.6) in general form and for this case $\{\hat{\varepsilon}^*\} = [\hat{\varepsilon}_1^* \quad \hat{\varepsilon}_2^*]^T$ as

$$\begin{Bmatrix} \hat{\varepsilon}_1^* \\ \hat{\varepsilon}_2^* \end{Bmatrix} = \begin{bmatrix} -1 & 0 & 1 \\ 0 & -1 & 1 \end{bmatrix} \begin{Bmatrix} _m\varepsilon_1^e \\ _m\varepsilon_2^e \\ _m\varepsilon_3^e \end{Bmatrix} = \begin{Bmatrix} _m\varepsilon_3^e - _m\varepsilon_1^e \\ _m\varepsilon_3^e - _m\varepsilon_1^e \end{Bmatrix}. \tag{2.5.7}$$

The most probable values of residual stresses are given by Eq. (2.5.8). The actual numerical and experimental results are represented in Section A.2.1 to A.2.5 of Appendix A.

2.7.2 Measurement of Three-Dimensional Residual Stresses Induced in Thick Plates

As measurement of three-dimensional residual stresses is difficult, it requires some considerable ingenuity. This process is explained in the following.

We assume the three bars of the model in Fig. 2.1 are arranged in the depth direction of the plate as shown in Fig. 2.2(b). Then, bar 1 is located on the top surface and bars 2 and 3 are inside. For measurement, we start with Fig. 2.7(a).

First, we work with the case where the measured values are free of error.

As the first step, we paste strain gauges on bar 1, although we cannot put strain gauges on bars 2 and 3, since they are supposed to be located inside of the plate.

Next, we cut bar 1 as indicated in Fig. 2.7(a) and measure the released strain generated by this cutting. The existed elastic strain of bar 1 is the opposite sign

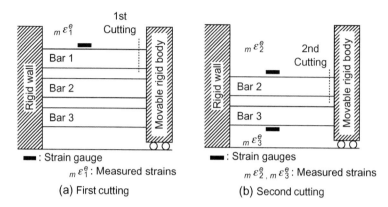

(a) First cutting

■ : Strain gauge
$_m\varepsilon_1^e$: Measured strains

(b) Second cutting

■ : Strain gauges
$_m\varepsilon_2^e$, $_m\varepsilon_3^e$: Measured strains

FIG. 2.7 Measuring procedure of three-dimensional residual stresses (thick-plate model).

of the measured released one, which is denoted by $_m\varepsilon_1^e$. This is expressed by the first equation of Eq. (2.4.6) as

$$_m\varepsilon_1^e = \frac{1}{3}[-2 \quad 1]\{\varepsilon^*\}, \tag{2.7.1}$$

where $\{\varepsilon^*\} = [\varepsilon_1^* \quad \varepsilon_2^*]^T$.

By this cutting, the effective inherent strain of bar 1, ε_1^*, has no relation with bars 2, and 3. Then the model consists of only bars 2, and 3, and the elastic strains left in bars 2 and 3, which are due only to ε_2^* of bar 2. The elastic strains for the renewed model can be evaluated according to the procedure indicated in Section 2.1 and expressed as

$$\begin{Bmatrix} \varepsilon_2^e \\ \varepsilon_3^e \end{Bmatrix} = \frac{1}{2} \begin{Bmatrix} -1 \\ 1 \end{Bmatrix} \varepsilon_2^*. \tag{2.7.2}$$

As bar 1 is removed, the surface of bar 2 is disclosed to put strain gauges on it. We cut bar 2 again as illustrated in Fig. 2.7(b) to release the stress (strain) on bar 2 completely. The existed elastic strain of bar 2 is the opposite sign of the measured one, which is denoted by $_m\varepsilon_2^e$. This is given from Eq. (2.7.2) as

$$_m\varepsilon_2^e = -\frac{1}{2}\varepsilon_2^*. \tag{2.7.3}$$

At the cutting of bar 2, bar 3 becomes free. Then, the elastic strain of bar 3 is relaxed, which is denoted as $_m\varepsilon_3^e$, and it is calculated from Eq. (2.7.2) as

$$_m\varepsilon_3^e = \frac{1}{2}\varepsilon_2^*. \tag{2.7.4}$$

A set of equations, (2.7.1) and (2.7.3), comprise the following elastic response equation, that is,

$$\{_m\varepsilon^e\} = [H^*]\{\varepsilon^*\}, \tag{2.7.5}$$

where

$$\{_m\varepsilon^e\} = [_m\varepsilon_1^e \quad _m\varepsilon_2^e]^T$$

$$[H^*] = \begin{bmatrix} -2/3 & 1/3 \\ 0 & -1/2 \end{bmatrix}$$

$$\{\varepsilon^*\} = [\varepsilon_1^* \quad \varepsilon_2^*]^T.$$

We can obtain the effective inherent strain $\{\varepsilon^*\}$ of the model by solving Eq. (2.7.5):

$$\{\varepsilon^*\} = [H^*]^{-1}\{_m\varepsilon^e\}, \tag{2.7.6}$$

where

$$[H^*]^{-1} = \begin{bmatrix} -3/2 & -1 \\ 0 & -2 \end{bmatrix}.$$

Here, $\{\varepsilon^*\}$ is the true inherent strain, and substitution of $\{\varepsilon^*\}$ into Eq. (2.4.5) provides the residual stresses in bars 1, 2, and 3 of the thick-plate model.

When experimental errors are inevitable in the measurement, we should estimate the most probable value of effective inherent strain, following the procedure demonstrated in Section 2.5.

To perform the standard procedure, the number of measured strains m should be more than that of the unknown inherent strains q. In the present case, if we measured the elastic strain of bar 3 at the cutting of bar 2, it would be given by Eq. (2.7.4). With this additional expression, the elastic response equation is composed by Eqs. (2.7.1) and (2.7.4) as

$$\{_m\varepsilon\} = [H^*]\{\varepsilon^*\} \tag{2.7.7.A}$$

where

$$[H^*] = \begin{bmatrix} -2/3 & 1/3 \\ 0 & -1/2 \\ 0 & 1/2 \end{bmatrix}.$$

This is expressed in terms of the elastic response matrix $[H^*]$. We can calculate the most probable value of the effective inherent strain $\{\hat{\varepsilon}^*\}$ by substituting $[H^*]$ into Eq. (2.5.6) as

$$\{\hat{\varepsilon}^*\} = ([H^*]^T[H^*])^{-1}[H^*]^T\{_m\varepsilon^e\}, \tag{2.7.7.B}$$

where

$$\{\hat{\varepsilon}^*\} = [\hat{\varepsilon}_1^* \quad \hat{\varepsilon}_2^*]^T$$

$$([H^*]^T[H^*])^{-1}[H^*]^T = \frac{1}{2}\begin{bmatrix} -3 & -1 & 1 \\ 0 & -2 & 2 \end{bmatrix}$$

$$\{_m\varepsilon^e\} = [_m\varepsilon_1^e \quad _m\varepsilon_2^e \quad _m\varepsilon_3^e]^T.$$

Taking a further step to insert $\{\hat{e}^*\}$ into the right side of Eq. (2.4.5), we can obtain the most probable value of the residual stress $\{\hat{\sigma}\}$.

In the preceding sections, the measuring methods of residual stresses have been demonstrated using simple models, utilizing inherent strains as parameters. In this procedure, we can also evaluate the unbiased estimate $\hat{s} = \sqrt{S/(m-q)}$ with the help of Eq. (2.5.9).

Some practical measurements were performed, and the results are represented in Sections A.3.2, A.4, and A.9.2 of Appendix A.

2.8 PREDICTION OF WELDING RESIDUAL STRESSES

It is often necessary to predict welding residual stresses of various types of joints without performing complex computation of thermal elastic-plastic analysis. In these cases, we can apply the following two simpler methods:

1. Using experimental databases; a very practical method. An experiment was performed on various types of welded joints and residual stresses were measured. The data were classified according to the type, size of the joint, and welding conditions so as to apply it to similar joints. The prediction is performed using this type of database.
2. Using effective inherent strain that is the source of residual stress. In the preceding sections, the relation between inherent strain and the resulting residual stress is described, and the inverse procedure to obtain the inherent strain from the measured strains was illustrated. Using these procedures, we can construct a database of inherent strains for a variety of welded joints from experiments, thermal elastic-plastic analysis, etc. Then we can predict the entire distribution of residual stresses of a specific type of welded joint by standard elastic calculation imposing the inherent strains as the source of residual stresses to a stress-free joint. As indicated in Section A.2.2 of Appendix A, the inherent strain distributions are similar for the same types of welded joints with a small variety of sizes, heat input, etc. Then, it is rather easy to adjust the distribution so as to apply the specimens with small differences in size of specimen, heat input, etc. This procedure can also be applied to predict welding deformation.

This chapter focused on the inherent strain method explained in detail using the idealized models. Practical examples of prediction of welding residual stresses and deformation are demonstrated in Sections 7.7 and 7.9, and also Sections A.2.2 and A.2.3 of Appendix A, respectively.

REFERENCES

[1] Watanabe S, Satoh K. Welding mechanics and its application. Tokyo: Asakura Shoten [in Japanese]; 1965.
[2] Ueda Y. Prediction and measuring methods of two- and three-dimensional welding residual stresses by using inherent strain as a parameter. In: Modeling in welding, hot powder forming, and casting. ASM International, Material Park, OH, USA, 1997 (ISBN:0-87170-616-4).

[3] Ueda Y, Fukuda K. New measuring method of three-dimensional residual stresses in long welded joints using inherent strains as parameters—Lz method. Trans of the ASME J Eng Mater Technol 1989;111:1–8.

[4] Ueda Y, Murakawa H, Ma NX. Measuring method for residual stresses in explosively clad plates and a method of residual stresses reduction. Trans ASME J Eng Mater Technol 1996;118:576–82.

[5] Ueda Y, Yuan MG. Prediction of residual stresses in butt welded plates using inherent strains. Trans ASME J Eng Mater Technol 1993;115:417–23.

[6] Yuan MG, Ueda Y. Prediction of residual stresses in welded T- and I-joints using inherent strains. Trans ASME J Eng Mater Technol 1996;118:229–34.

[7] Ueda Y, Fukuda K, Nakacho K, Endo M. A new measuring method of residual stresses with the aid of finite element method and reliability of estimated values. Trans JWRI 1975;4 (2):123–31.

[8] Ueda Y. Sectioning methods. In: Handbook of measurement of residual stresses. Lilburn, GA: Fairmont Press, Society for Experimental Mechanics; 1996.

[9] Ueda Y, Ma N. A function method for estimating inherent strain distributions. Trans JWRI 1994;23(1):71–8.

[10] Ma N, Ueda Y. TLyLz-method and T-method for measuring 3-dimensional residual stresses in bead-on-plate welds. Trans JWRI 1994;23(2):239–47.

[11] Ueda Y, Nakacho K, Kim Y, Murakawa H. Fundamentals of thermal-elastic-plastic-creep analysis and measurement of welding residual stress for numerical analysis. Measurement of welding residual stress. J JWS 1986;55(8):458–65 [in Japanese].

Chapter 3

Mechanical Simulation of Welding

As discussed in Chapter 1, fusion bonding such as arc welding is a process of joining parts by melting them. In this process, concentrated heat is applied to the area to be joined. Due to such concentrated heat, the region near the joint experiences high temperature and a large temperature gradient. This causes welding distortion and residual stress. The welding distortion is one of the major causes of the dimensional inaccuracy of products, and it can be an obstacle to introducing robot welding when it exceeds the allowable limit. Welding residual stress has a significant influence on buckling and ultimate strength, the strength against the brittle fracture, fatigue, and stress corrosion cracking of structures [1–6]. Therefore, accurate estimation of the distortion, strain, and residual stress induced by welding is necessary to assess the strength and the reliability of structures. Theoretical methods such as the finite element method (FEM) are powerful tools for their prediction and to understand how they are formed in the welding process.

In this chapter, the basic theories of the computation of welding distortion and residual stress are explained using very simple one-dimensional models. The ideas explained for the one-dimensional problem are general so they can be readily extended to two-dimensional or three-dimensional problems. These details are included on the companion website of this book for those interested in studying them.

3.1 HEAT FLOW AND TEMPERATURE DURING WELDING

In the welding process, metals such as steel are melted by heat that is generated by a moving arc. After the arc passes, the molten metal is cooled by the dissipation of the heat, and it solidifies and shrinks by thermal contraction. Since residual stress and distortion are produced by the thermal contraction, it is necessary to know the time history and distribution of temperature for their theoretical estimation.

In this section, heat conduction and heat transfer, the two fundamental forms of heat flow or dissipation will be explained intuitively using the three-cylinder model, which is similar to the three-bar model discussed in Chapters 1 and 2.

Welding Deformation and Residual Stress Prevention
© 2012 Elsevier Inc. All rights reserved.

The flow of heat is controlled by material properties such as thermal conductivity and specific heat. The variation of these material constants is shown using steel and aluminum as examples. Thermal and mechanical material constants also change with temperature. In steel, yield stress becomes almost zero when the temperature reaches around 750°C. Such a temperature is called the *mechanical melting point*, because at this point the material loses its resistance against deformation just like liquid. The mechanical melting point is a very important index for understanding the process that produces welding distortion and residual stress. As will be explained in this chapter, the size of the area in which the temperature exceeds the mechanical melting point is one of the representative scales which indicates the size of the mesh to be used for the finite element analysis.

In the last part of this chapter, how to compute heat flow in welding, which changes with time, is explained using the three-cylinder model.

3.1.1 Heat Supply, Diffusion, and Dissipation

The welding process is divided into the following three stages with respect to temperature changing with time:

1. Heating of metal by given heat
2. Melting of metal
3. Cooling of metal and solidification

In the heating process, part of the electric power P (J/s), which is the product of the current I (A) and voltage U (V), is given through the radiation from the arc, the molten weld wire and Joule heat form the weld metal. The remaining part of the power is dissipated into the atmosphere through the radiation and convection, as shown in Fig. 3.1. Therefore, the net power actually given to the weld joint P_{net} is given by

$$P_{net} = \eta P = \eta I U \ (\text{J/s}), \qquad (3.1.1)$$

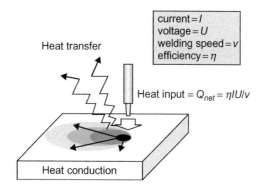

FIG. 3.1   Heat flow in welding.

where η is the efficiency. In submerged arc welding, the efficiency is high. It is about 0.9–0.99. The efficiency of TIG (Tungsten Inert-Gas) welding is low, about 0.3–0.6 [2]. To discuss the change of temperature with time it is convenient to use the heat input per unit weld length Q instead of the power. When the welding speed is v (mm/s), the net heat input per unit length of weld Q_{net} becomes

$$Q_{net} = \eta P/v = \eta IU/v \ \ (\text{J/mm}). \tag{3.1.2}$$

The base metal and the weld metal are heated by this heat input Q_{net}. If the heat is given to the metal with cross-sectional area s (mm^2) and assuming that the heat does not move, the average temperature increase T_{av} is given by

$$T_{av} = Q_{net}/(c\rho s), \tag{3.1.3}$$

where c and ρ are, respectively, density and specific heat. Though the equation is simple, it tells us the following basic ideas:

1. Temperature increase T_{av} is proportional to the heat input Q_{net}.
2. For the same heat input Q_{net}, the temperature increase of the metal is small when $c\rho$ is large.
3. By combining Eq. (3.1.2), which gives Q_{net}, and Eq. (3.1.3), the following relation is obtained:

$$T_{av} = \eta IU/(c\rho sv) \tag{3.1.4}$$

This equation tells us that, for the same current I and voltage U, the temperature increase T_{av} becomes small when the welding speed is large.

With the approach of the arc, the source of the heat supply, the temperature increases then it decreases after the pass of the heat source because the heat given to the weld metal is spread with time. There are two ways of spreading or dissipation of heat. One is heat conduction, which is the spread of heat in the metal. Another is heat transfer, which is the dissipation of heat into the atmosphere.

In heat conduction, the flow of heat is proportional to the thermal conductivity λ (J/smm°C), which is an index showing how well the heat flows in the material with the temperature gradient. This relation is known as Fourier law [7–10]. The amount of heat flow becomes large when the cross-sectional area and the time are large, that is,

heat flow in unit time $\propto \lambda \times$ temperature gradient $\times$ cross-sectional area. (3.1.5)

This relation can be described in a more theoretical way. The heat flows through the metal with unit cross-section (1 mm^2) in unit time (1 sec) and $\dot{q}_c$ can be given by

$$\dot{q}_c = \lambda \times \text{temperature gradient.} \tag{3.1.6}$$

In general, thermal conductivity λ changes with temperature. If the material shows no phase transformation accompanying the evolution of the microstructure, the thermal conductivity λ can be roughly described as a linear function of temperature. For example, the thermal conductivity of carbon steel at 800°C becomes almost half of that at room temperature.

Heat transfer is the phenomenon in which heat flows out or across the surface of the material. In heat transfer, two forms of heat flow are involved. The first form is the heat that moves from the metal surface to the surrounding fluid, such as the air, and is carried away by the convection of the fluid. The second form is the radiation where the heat is emitted in the form of light. When the metal is in the fluid, the radiation from the fluid to the metal surface needs to be considered in some cases.

Newton's law is applied to the convection. The heat flow from the metal surface to the atmosphere is proportional to the product of difference between the temperature of the atmosphere and that of the metal surface $(t - t_0)$ and the heat transfer coefficient β_c (J/smm^2°C), which is a material constant that shows how easily the heat moves across the metal surface.

$$\text{heat transfered to atmosphere in unit time} \propto \beta_c(T - T_0). \tag{3.1.7}$$

The heat transfer coefficient β_c changes with the condition of the fluid. In steel in static air, for example, its value is in the range

$$\beta_c = 3 \sim 30 \, \text{J/(sm}^2\text{°C)}. \tag{3.1.8}$$

If the air is moving with the wind, the heat transfer coefficient becomes one order higher.

The heat flow from the metal surface to the atmosphere by radiation is given by Stefan-Boltzman's law, and the amount of heat going out from the metal surface is proportional to the difference between the fourth power of the temperature on the surface and that of the atmosphere.

Heat flowing into the atmosphere through radiation in unit time is

$$\propto \varepsilon C_0 \left\{ (T + 273)^4 - (T_0 + 273)^4 \right\}, \tag{3.1.9}$$

where C_0 is the radiation constant for the black body (the Stefan-Boltzmann constant) and its value is 5.67×10^{-8} (J/m^{2}s°C) [5, 8]. The parameter ε is the emissivity of the material that changes with the surface condition of the material. In a steel surface covered by oxide, $\varepsilon \cong 0.8 - 0.9$. If the surface is polished, $\varepsilon \cong 0.4$.

In heat transfer in welding, both convection and radiation occur simultaneously, and the heat loss by heat transfer becomes the sum of Eq. (3.1.7) and Eq. (3.1.9). For convenience, it can be described in the same form as Newton's law:

Heat flowing into the atmosphere through heat transfer is

$$\propto \beta(T - T_0), \tag{3.1.10}$$

where β is defined by the following equation, and also shown in Fig. 3.2:

$$\beta = \beta_c + \varepsilon C_0 \left\{ (T + 273) + (T_0 + 273) \right\} \left\{ (T + 273)^2 + (T_0 + 273)^2 \right\} \, \text{J/m}^2\text{s°C}. \tag{3.1.11}$$

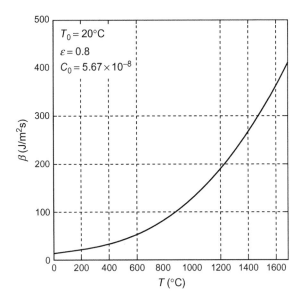

FIG. 3.2 Heat transfer coefficient considering convection and radiation.

More correctly, the heat flows out from the surface to the atmosphere through heat transfer per unit area and unit time $\dot{q}_t$ is given by the equation

$$\dot{q}_t = \beta(T - T_0). \tag{3.1.12}$$

As seen from Eq. (3.1.11), the heat transfer coefficient β changes with temperature. In the low-temperature range, the convection is dominant while the radiation becomes dominant when the temperature is high.

We have seen that the thermal conductivity λ and the heat transfer coefficient β are functions of temperature. Similarly, the specific heat c and density ρ also change with temperature.

3.1.2 Simple Heat Flow Model

As we know intuitively, heat flows from high to low just as water flows. If we have two bodies connected to each other, the heat flows from the body with the higher temperature to the body with the lower temperature. Based on this, we can replace the heat flow in the thickness direction under welding with the water flow in three cylinders connected with tubes. Figure 3.3 shows the through-thickness temperature distribution. The coordinate z is taken upward with its origin on the bottom surface. The welding is assumed to be performed on the top surface. The plate is divided into three parts from the top surface; namely, A, B, and C in the thickness direction. The welding heat is given to part A. The flow of the heat in the thickness direction can be replaced by the water flow among three cylinders connected with tubes, as shown in Fig. 3.4.

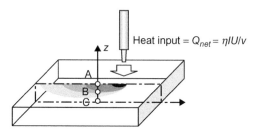

FIG. 3.3   Through-thickness temperature distribution.

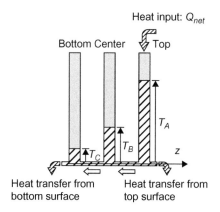

FIG. 3.4   Three-cylinder model.

For simplicity, the heat flows mostly in the thickness direction, and the flow in the length or the width direction can be neglected. The welding heat Q_{net} corresponds to the water poured into cylinder A. The water flowing out from the hole opened at the right and left sides represents the heat flowing out through the heat transfer. The water flowing through the tubes represents the heat flow due to the heat conduction. Then the height of the water in the cylinders T_A, T_B, and T_C represent the temperature at the top surface, center of thickness, and bottom surface.

Using this very simple model, we can intuitively understand fundamental phenomena in heat flow associated with welding.

3.1.2.1 Change of Temperature Distribution with Time (without Heat Transfer)

Figure 3.5 illustrates how the temperature on the top surface rises by the welding heat given to the top surface. It also illustrates how the temperatures in the center and on the bottom surface rise with time due to the heat flow through conduction. Since heat is not allowed to flow out from the surface through

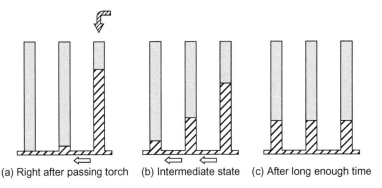

(a) Right after passing torch   (b) Intermediate state   (c) After long enough time

FIG. 3.5 Through-thickness temperature distribution changing with time (without heat transfer).

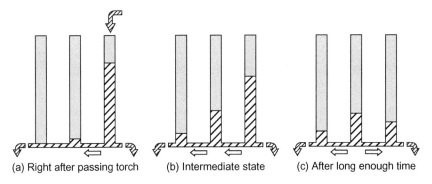

(a) Right after passing torch   (b) Intermediate state   (c) After long enough time

FIG. 3.6 Through-thickness temperature distribution changing with time (with heat transfer).

heat transfer in this case, the temperature becomes uniform through the thickness after sufficient time, as shown in Fig. 3.5(c).

3.1.2.2 Change of Temperature Distribution with Time (with Heat Transfer)

The change of temperature distribution with time when the heat transfer is considered is shown in Fig. 3.6. As opposed to the case without heat transfer, the temperature cools down to room temperature after a sufficiently long time because all heat flows out from the plate through heat transfer. Before the complete cool down, the temperatures in the mid-thickness becomes higher than on the surfaces, as shown in Fig. 3.6(c).

3.1.2.3 Influence of Heat Input

The influence of heat input is illustrated in Fig. 3.7. As illustrated, the temperature becomes high when the heat input is large.

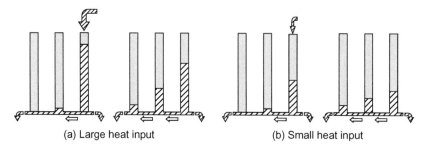

(a) Large heat input   (b) Small heat input

FIG. 3.7   Influence of heat input on through-thickness temperature distribution.

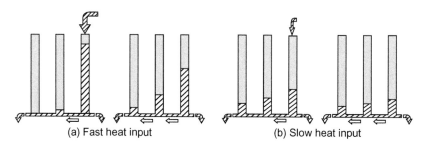

(a) Fast heat input   (b) Slow heat input

FIG. 3.8   Influence of heat input rate on through-thickness temperature distribution.

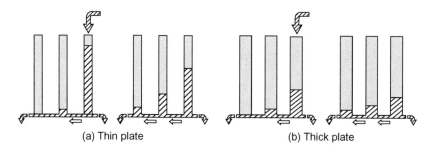

(a) Thin plate   (b) Thick plate

FIG. 3.9   Influence of plate thickness on through-thickness temperature distribution.

3.1.2.4 Influence of Speed of Heat Input

Figure 3.8 shows the comparison of the cases with the same heat input per unit length but with different powers. If the power is large and heat is given in a short time, the temperature becomes high. If the power is small, the temperature becomes low. This explains why high power is required to melt metal by welding.

3.1.2.5 Influence of Plate Thickness

Figure 3.9 shows the influence of plate thickness when welding conditions such as heat input and welding speed are the same. The thin plate corresponds to the

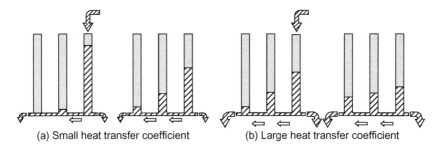

(a) Small heat transfer coefficient   (b) Large heat transfer coefficient

FIG. 3.10 Influence of heat transfer on through-thickness temperature distribution.

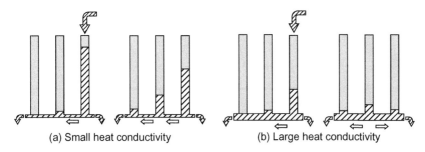

(a) Small heat conductivity   (b) Large heat conductivity

FIG. 3.11 Influence of heat conductivity on through-thickness temperature distribution.

thin cylinder. The temperature, the height of the water, becomes higher compared to the thick plate when the same heat input is given.

3.1.2.6 Influence of Heat Conduction

Materials with high heat conductivity correspond to cylinders connected to thick tubes, as shown in Fig. 3.10. When the same amount of water is poured, the temperature becomes lower compared to the case with low heat conductivity because heat flows quickly through conduction. The through-thickness temperature gradient also becomes small when the heat conductivity is high.

3.1.2.7 Influence of Heat Transfer

The influence of heat transfer is illustrated in Fig. 3.11. When the heat transfer coefficient is large, the temperature decreases quickly because the heat is dissipated to the atmosphere faster.

3.1.3 Differences in Material Properties

As discussed in the preceding section, heat flow is influenced by material properties such as heat conductivity, specific heat, and density. The values of these properties change for different materials. Table 3.1 shows the comparison of

TABLE 3.1 Material Properties of Aluminum and Mild Steel

	Aluminum	Mild Steel	Alumi./Steel
Mechanical melting temperature: T_m (°C)	660	1520	0.43
Density: ρ (kg/m³)	2.7×10^3	7.86×10^3	0.34
Specific heat: c (J/kg·°C)	9.2×10^2	4.6×10^2	2
Young's modulus: E (GPa)	70.6	210	0.34
Thermal expansion ratio: α (1/°C)	2.4×10^{-5}	1.2×10^{-5}	2
Thermal conductivity: λ (J/ms°C)	217	50.2	4.3
Thermal diffusivity: $\kappa = \lambda/c\rho$ (m²/s)	8.73×10^{-5}	1.39×10^{-5}	6.3
Latent heat: q (J/kg)	4.05×10^5	2.76×10^5	1.5
$1/c\rho$	4.02×10^{-7}	2.77×10^{-7}	1.5
$\alpha/c\rho$	9.66×10^{-12}	3.32×10^{-12}	2.9

aluminum and mild steel at room temperature. As indicated by the ratio in the right column, significant differences are observed in thermal conductivity and thermal diffusivity. Their values for aluminum are 4.3 times and 6.3 times higher than those for mild steel. This means heat can flow much faster in aluminum. The thermal expansion ratio α and the parameter $\alpha/c\rho$, which influence welding distortion and residual stress, are 2 times and 2.9 times higher than those of mild steel. Larger thermal conductivity and large thermal expansion ratio correspond to the fact that the welding of aluminum is more difficult and produces larger distortion than mild steel.

3.1.4 Change of Material Properties with Temperature

In the welding of steel, the material is heated up to roughly 2000°C and cooled down to room temperature. The material properties, including heat conductivity, specific heat, density, heat transfer coefficient, thermal expansion ratio, Young's modulus, yield stress and tensile strength, change with temperature. Figures 3.12 to 3.18 show the variations of thermal and mechanical properties with temperature [11–15]. Among these, the specific heat becomes large at around 800°C in Fig. 3.12 due to the latent heat associated with the phase transformation. A similar phenomenon is observed in the thermal expansion ratio, which is also due to phase transformation. The yield stress becomes very small at high temperatures over 600–700°C. Yield stress close to zero means that the material loses its resistance to deformation or behaves like fluid. For this reason, this temperature is called the mechanical melting temperature.

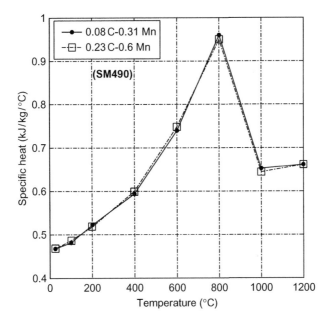

FIG. 3.12 Specific heat [10].

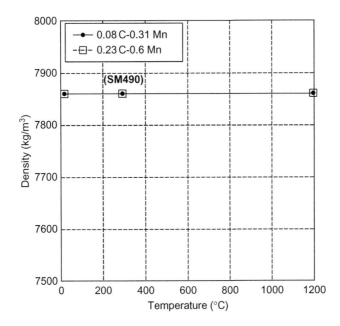

FIG. 3.13 Density [10].

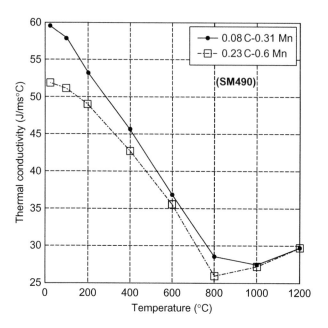

FIG. 3.14   Thermal conductivity [10].

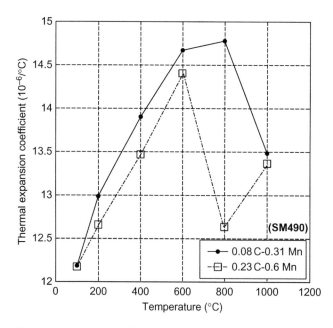

FIG. 3.15   Thermal expansion ratio [10].

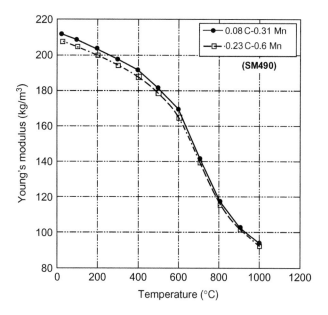

FIG. 3.16   Young's modulus [11,12].

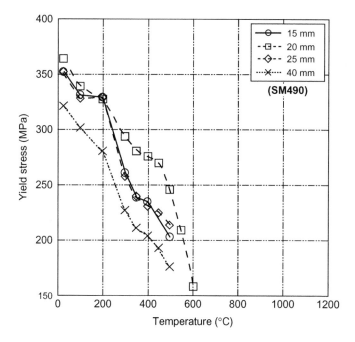

FIG. 3.17   Yield stress [13].

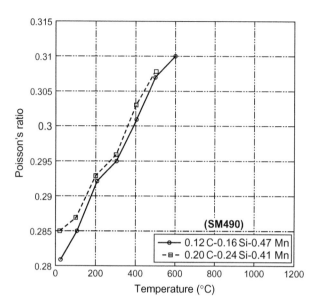

FIG. 3.18   Poisson's ratio [11,14].

As previously mentioned, one of the reasons for the changes in the thermal and mechanical properties with temperature is phase transformation. Figure 3.19 is the phase diagram of carbon steel [15], which shows the relation between the stable phase and temperature. In this figure, A_1 (roughly 723°C) is the temperature at which the eutectoid transformation starts. A_3 (roughly 723–910°C depending on the carbon contents) is the temperature at which the phase transformation to Austenite ends during heating. The metal heated over A_1 will have the solid phase transformation from Austenite to Martensite or other phases according to the cooling rate. Austenite is the soft phase with low Young's modulus and yield stress, while Martensite is the hard phase. As shown in the figure, the stable phase changes with the content of carbon and other elements. If the carbon content is 0.008–2%, the material is steel. It is iron if the carbon content is less than 0.008–2%, and it is cast iron if it is larger.

Phase transformation can also occur in welding in the region near the weld metal where the temperature is high. The base metal heated above A_1 is called the heat-affected zone (HAZ).

As shown in Fig. 3.19, Austenite is a stable phase when the temperature is high, and Ferrite is stable when the temperature is low. When Austenite is cooled at a high cooling rate, Martensite forms. Because Martensite is hard and brittle, it may cause welding cracking and low toughness. The relation between the cooling rate and the resulting phase can be predicted using the continuous cooling transformation (CCT) diagram. Figure 3.20 shows the CCT [4] of 500 MPa–class high tensile strength steel as an example. In the figure, symbols a, f and m indicate

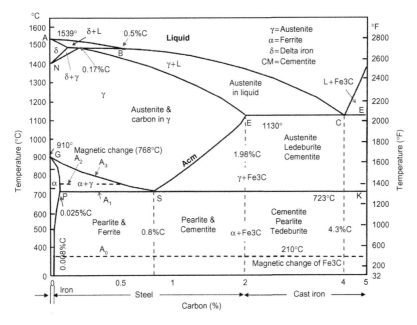

FIG. 3.19 Phase diagram of carbon steel [16].

Austenite, Ferrite and Martensite, respectively. The cooling lines z, f, p, and e represent the critical cooling condition at which the resulting phase changes. For example, when the cooling rate is higher than curve z, all the material becomes Martensite. As the cooling rate becomes lower than z, the fraction of Martensite decreases and the hardness becomes low. As a practical parameter to indicate the speed of cooling, the time in which the temperature decreases from 800°C to 500°C is commonly used.

3.1.5 Characteristic Temperature and Length

As a consequence of the flow and dissipation of heat introduced by welding through heat conduction and heat transfer, temperature distribution changes with time. After a long enough time, the temperature in the whole welded structure returns to room temperature T_0. To understand the process in which stress and deformation are produced during the welding thermal cycle, there are temperatures, such as the metallurgical melting temperature, the mechanical melting temperature T_m, and the yield temperature T_Y, which characterize the phenomena. By checking when and where these characteristic temperatures are passed during the heating and cooling stage, the progress of the phenomena can be easily understood.

Maximum temperature: 1300°C

C_e', C_p', C_f', C_z': Critical cooling time

a: Austenite, f: Ferrite, p: pearlite

m: martensite, Z_w: Bainite (middle phase)

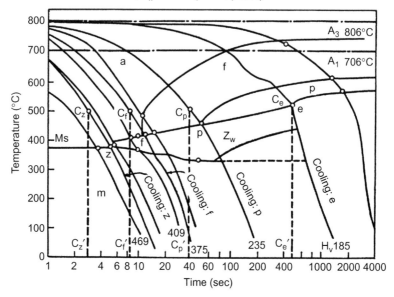

FIG. 3.20 Continuous cooling transformation diagram of 500 MPa-class steel [4].

3.1.5.1 Metallurgical Melting Temperature

The material heated above the metallurgical melting temperature can be regarded as the molten metal in the weld pool. By comparing the computed penetration shape and the measured shape, we can judge whether the heat input distribution and the heat input Q_{net} are correctly given in the simulation.

3.1.5.2 Mechanical Melting Point

As already discussed, when the temperature is higher than the mechanical melting temperature, the deformation of the metal becomes plastic deformation dominant. Since the welding residual stress and the distortion are produced by an irreversible strain component such as the plastic strain, the area heated above the mechanical melting temperature is very important. Specifically, the transverse shrinkage and the transverse bending (angular distortion) are mostly produced in this temperature range.

3.1.5.3 Yield Temperature

In contrast to the fact that the transverse shrinkage and the angular distortion are produced at a high temperature over the mechanical melting temperature, the longitudinal shrinkage and the residual stress are produced at a relatively low temperature above the yield temperature. The yield temperature T_Y here is the temperature at which the material fully restraint in one direction reaches the yield stress σ_Y by thermal expansion. Thus, it is defined by

$$T_Y = \sigma_Y/(\alpha E), \tag{3.1.13}$$

where α and E are the thermal expansion ratio and the Young's modulus. For simplicity, room temperature is assumed to be zero. In the case of mild steel, by substituting $\sigma_Y = 240\,\text{MPa}$ and $\alpha = 1.2 \times 10^{-5}\,1/°C$ into Eq. (3.1.13), the yield temperature T_Y becomes 100°C. The reason the shrinkage and the bending in the transverse direction are related to the mechanical melting temperature and those in the longitudinal direction are related to the yield temperature comes from the large restraint in the welding direction. The restraint in the welding direction is very strong, while that in the transverse direction is small.

3.1.5.4 Highest Temperature Reached

When the heat input is Q_{net} and the initial temperature is T_0, the highest temperature experienced T_{max} at the point distance r away from the weld center during the thermal cycle for thin and thick plates is given approximately by the following equations [7]. As illustrated in Fig. 3.21, the equation for the thin plate is derived based on the assumption that the welding speed is fast enough so that the welding heat can be idealized as a instantaneous heat source distributing along the welding line, and the heat flows only in the width direction. In a thick plate, the welding heat is applied as the instantaneous line heat source, and the heat is assumed to flow in the radial direction with the welding line at its center.

$$\text{In a thin plate: } T_{max} - T_0 = \frac{1}{\sqrt{2\pi e}} \frac{Q_{net}}{c\rho h} \frac{1}{r} = 0.242 \frac{Q_{net}}{c\rho h} \frac{1}{r} \tag{3.1.14}$$

$$\text{In a thick plate: } T_{max} - T_0 = \frac{2}{\pi e} \frac{Q_{net}}{c\rho} \frac{1}{r^2} = 0.234 \frac{Q_{net}}{c\rho} \frac{1}{r^2} \tag{3.1.15}$$

(a) Thin plate   (b) Thick plate

FIG. 3.21 Heat flow in thin and thick plates.

If the initial temperature T_0 is zero for simplicity, the distance from the weld center r at which the highest temperature is T_{max} is given by the following.

$$\text{In a thin plate: } r = \frac{1}{\sqrt{2\pi e}} \frac{Q_{net}}{c\rho h} \frac{1}{T_{max}} = 0.242 \frac{Q_{net}}{c\rho h} \frac{1}{T_{max}} \qquad (3.1.16)$$

$$\text{In a thick plate: } r^2 = \frac{2}{\pi e} \frac{Q_{net}}{c\rho} \frac{1}{T_{max}} = 0.234 \frac{Q_{net}}{c\rho} \frac{1}{T_{max}} \qquad (3.1.17)$$

Using these equations, the sizes of the zone in which the temperature is higher than the metallurgical melting temperature, the mechanical melting temperature and the yield temperature can be estimated if the specific heat c, the density ρ, the thickness of the plate h, and the heat input Q_{net} are given. For example, if the plate is 5 mm-thick mild steel, the heat input is 1 kJ/mm and the metallurgical melting temperature, the mechanical melting temperature, and the yield temperature are 1450°C, 700°C, and 200°C, respectively. The corresponding values of r for the thin and thick plates are summarized in Table 3.2. As seen from the table, the sizes of the metallurgical melting zone, the mechanical melting zone, and the yielding zone for the thick plate are much smaller than those estimated for the thin plate. In general, the yield zone is directly related to the longitudinal shrinkage and the longitudinal residual stress, while the mechanical melting zone is closely related to the transverse shrinkage. In a thick plate, the former is roughly twice as large as the latter. In a thin plate, the size of the yield zone relative to the mechanical melting point becomes larger. The cases with different plate thicknesses are summarized in Fig. 3.22. When the thickness is larger than 7 mm, the size of the metallurgical melting zone estimated as the thin plate becomes smaller than the value estimated as the thick plate. Also, when the thickness is larger than 10 mm, the size of the mechanical melting estimated as the thin plate becomes smaller than the value estimated as the thick plate. These conflict with the common idea that the size of the temperature zone is larger in a thin plate than a thick plate. This means that when the heat input is 1 kJ/mm and the thickness is larger than 10 mm, the equation for the thick plate must be used rather than that for the thin plate to estimate the size of the mechanical melting zone. Otherwise, the size of the mechanical melting zone will be underestimated.

TABLE 3.2 Characteristic Length for Welding with Q_{net} = 1 kJ/mm, T_0 = 0

	Size of Metallurgical Melting Zone	Size of Mechanical Melting Zone	Size of Yield Zone
Thin plate (h = 5 mm) Eq. (3.1.16)	9.23 mm	19.1 mm	66.9 mm
Thick plate ($h = \infty$) Eq. (3.1.17)	6.68 mm	9.62 mm	18 mm

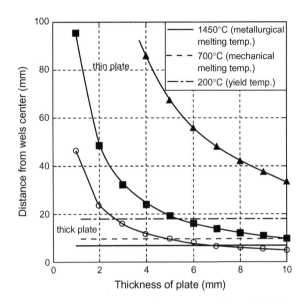

FIG. 3.22 Comparison of characteristic lengths estimated for thin and thick plates.

3.1.5.5 Cooling Speed

The cooling rate $\dot{T}$ at the center of the weld when the temperature cools down to $T°C$ can be estimated using the following equations. Since the cooling rate is closely related to the hardness and the grain size, which influence the toughness of the material, it is an important information for welding along with the highest temperature.

In a thin plate: $\dot{T} = 2\pi k \left(\dfrac{c\rho h}{Q_{net}}\right)^2 (T - T_0)^3 = 2\pi c\rho\lambda \left(\dfrac{h}{Q_{net}}\right)^2 (T - T_0)^3$ (3.1.18)

In a thick plate: $\dot{T} = 4\pi k \left(\dfrac{c\rho}{Q_{net}}\right)(T - T_0)^2 = 4\pi \dfrac{\lambda}{Q_{net}} (T - T_0)^2.$ (3.1.19)

From these equations, it is seen that the cooling rate R becomes large when the thermal conductivity λ is large and the heat input Q_{net} is small. In a thin plate, it becomes large when the thickness h is large.

3.1.5.6 Resolution Required for Computation

As discussed, the size and shape of the penetration and the area in which the plastic strain is formed are determined by the temperature relative to the characteristic temperatures. The size of the area where the temperature is higher than the characteristic temperature gives the reference to determine the mesh size in the finite element simulation. Depending on the purpose of the simulation, the appropriate mesh size can be determined. For example, to obtain the shape of the penetration accurately, the mesh size near the weld pool must be smaller than a quarter of the size of the area where the temperature is higher than the metallurgical melting temperature. To estimate the welding residual stress in the welding direction, the reference size is the area where the temperature is higher than the yield temperature. The area away from these areas can be divided using a larger mesh.

3.1.6 Simple Method for Solving the Heat Conduction Problem

The method for solving the heat conduction problem is explained using the simple three-cylinder model shown in Section 3.1.2. The heat flow in the welding is simplified, as shown in Fig. 3.23. The heat flows in the thickness direction within the rectangular parallelepiped consisting of cells A, B, and C. The thickness of each cell is one-third of the plate thickness h. Denoting the temperature in cells A, B, and C as T_A, T_B, and T_C and the cross-sectional area by s, the heat accumulated in each cell Q_A, Q_B, and Q_C is given by

$$Q_A = (c\rho sh/3)T_A$$
$$Q_B = (c\rho sh/3)T_B \qquad (3.1.20)$$
$$Q_C = (c\rho sh/3)T_C,$$

where c and ρ are the specific heat and density. The heat $\dot{q}_{AB}$ flowing from A to B and the heat $\dot{q}_{BC}$ flowing from B to C in a unit time are given by

$$\dot{q}_{AB} = \lambda s(T_A - T_B)(3/h)$$
$$\dot{q}_{BC} = \lambda s(T_B - T_C)(3/h), \qquad (3.1.21)$$

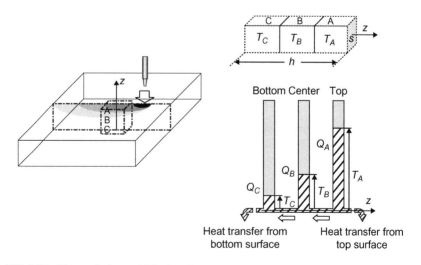

FIG. 3.23 Three-cylinder model for heat flow in thickness direction.

where λ is the thermal conductivity. In the same way, the heat $\dot{q}_A$ dissipating into the atmosphere from the top surface through the heat transfer and the heat $\dot{q}_C$ dissipating from the bottom surface in a unit time are given by

$$\dot{q}_A = \beta s(T_A - T_0)$$
$$\dot{q}_C = \beta s(T_C - T_0),$$

(3.1.22)

where β is the heat transfer coefficient considering the convection and the radiation. Then, the total heats $\dot{Q}_A$, $\dot{Q}_B$, and $\dot{Q}_C$ flowing into A, B, and C in a unit time are given by

$$\dot{Q}_A = -(3\lambda s/h)(T_A - T_B) - \beta s(T_A - T_0)$$
$$\dot{Q}_B = -(3\lambda s/h)(2T_B - T_A - T_C)$$
$$\dot{Q}_C = -(3\lambda s/h)(T_B - T_C) - \beta s(T_C - T_0).$$

(3.1.23)

Since the temperatures in cells A, B, and C change with these heats, the time rates of the temperatures dT_A/dt, dT_B/dt, and dT_C/dt are given by

$$(c\rho sh/3)\frac{dT_A}{dt} = -(3\lambda s/h)(T_A - T_B) - \beta s(T_A - T_0)$$
$$(c\rho sh/3)\frac{dT_B}{dt} = -(3\lambda s/h)(2T_B - T_A - T_C)$$
$$(c\rho sh/3)\frac{dT_C}{dt} = -(3\lambda s/h)(T_C - T_B) - \beta s(T_C - T_0).$$

(3.1.24)

Rearranging these equations,

$$(c\rho h/3)\frac{dT_A}{dt} + (3\lambda/h)(T_A - T_B) + \beta T_A = \beta T_0$$
$$(c\rho h/3)\frac{dT_B}{dt} + (3\lambda/h)(2T_B - T_A - T_C) = 0$$
$$(c\rho h/3)\frac{dT_C}{dt} + (3\lambda/h)(T_C - T_B) + \beta T_C = \beta T_0.$$

(3.1.25)

The above equations can be described in matrix form as

$$[C]\left\{\frac{dT}{dt}\right\} + [K]\{T\} = \{Q\},$$

(3.1.26)

where,

$$\left\{\frac{dT}{dt}\right\} = \left\{\begin{array}{c} dT_A/dt \\ dT_B/dt \\ dT_C/dt \end{array}\right\}, \quad \{T\} = \left\{\begin{array}{c} T_A \\ T_B \\ T_C \end{array}\right\}$$

(3.1.27)

$$[C] = (c\rho sh/3)\begin{bmatrix} 1 & 0 & 0 \\ 0 & 1 & 0 \\ 0 & 0 & 1 \end{bmatrix} \quad \text{(heat capacity matrix)}$$

(3.1.28)

$$[K] = (3\lambda s/h) \begin{bmatrix} 1 & -1 & 0 \\ -1 & 2 & -1 \\ 0 & -1 & 1 \end{bmatrix} + \beta s \begin{bmatrix} 1 & 0 & 0 \\ 0 & 0 & 0 \\ 0 & 0 & 1 \end{bmatrix}$$

(3.1.29)

(heat conduction matrix)   (heat transfer matrix)

$$[Q] = \beta s T_0 \begin{Bmatrix} 1 \\ 0 \\ 1 \end{Bmatrix}$$

(3.1.30)

The temperature distribution and its change with time can be obtained by solving Eq. (3.1.26). As it is seen intuitively, Eq. (3.1.26), which involves two unknown variables $\{dT/dt\}$ and $\{T\}$, cannot be solved. This equation can be solved numerically by introducing a finite difference method such as the Crank-Nicolson method or the central difference. Using the central difference, the temperature $T(t + \Delta t)$ and its time rate $dT(t + \Delta t)/dt$ at time $t + \Delta t$ are approximated by

$$T(t + \Delta t/2)\} = \frac{1}{2} \left(\{T(t + \Delta t)\} + \{T(t)\} \right)$$
$$\frac{dT(t + \Delta t/2)}{dt}\} = \frac{\{T(t + \Delta t)\} - \{T(t)\}}{\Delta t}.$$

(3.1.31)

By substituting these equations into Eq. (3.1.26) for time $(t + \Delta t/2)$, the following equation is derived:

$$\left(\frac{1}{2}[K] + \frac{1}{\Delta t}[C] \right) \{T(t + \Delta t)\} = \left(-\frac{1}{2}[K] + \frac{1}{\Delta t}[C] \right) \{T(t)\} + \{Q\}.$$ (3.1.32)

Since $\{T(t)\}$ on the right side is the known temperature in the past, Eq. (3.1.32) can be solved for unknown temperatures $\{T(t + \Delta t)\}$. The explanation here may be too advanced for beginners. For those who want to simulate the temperature distribution in welding using the FEM, it is enough to understand that the temperature distribution during welding can be computed by splitting time into small time increments Δt and solving Eq. (3.1.32) step by step.

If large time increment is used, the number of times to solve Eq. (3.1.32) can be reduced and computational time becomes short. However, there is a limit on the size of the time increment Δt, and such a limit can be easily understood. Let's consider the simple model shown in Fig. 3.24. This model consists of two cells connected to each other. Heat can flow between cells A and B through heat conduction. For simplicity, heat dissipation through heat transfer is not considered.

From Eq. (3.1.21), the heat flowing from A to B in the time increment Δt is given as

$$\Delta t q_{AB} = \Delta t \lambda s (T_A - T_B)(2/h),$$

(3.1.33)

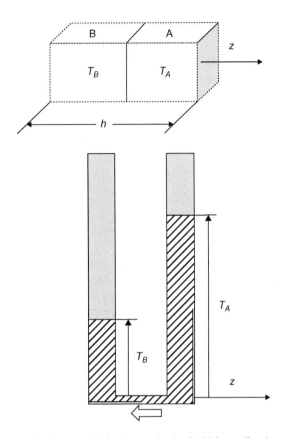

FIG. 3.24 Connected cylinder model for heat conduction in thickness direction.

while the changes of temperature ΔT_A and ΔT_B due to this heat flow are given by

$$\Delta T_A = -\Delta t q_{AB}/(c\rho sh/2) = -\Delta t \lambda (T_A - T_B)/(c\rho h^2/4)$$
$$\Delta T_B = \Delta t q_{AB}/(c\rho sh/2) = \Delta t \lambda (T_A - T_B)/(c\rho h^2/4). \tag{3.1.34}$$

For the heat to flow from A to B at time $\Delta t + t$, $T_A(t + \Delta t) > T_B(t + \Delta t)$ must hold, that is,

$$T_A + \Delta T_A - T_B - \Delta T_B = (T_A - T_B)\left(1 - \Delta t \lambda /(c\rho h^2/4)\right) > 0, \tag{3.1.35}$$

or,

$$\Delta t < (c\rho/\lambda)(h/2)^2. \tag{3.1.36}$$

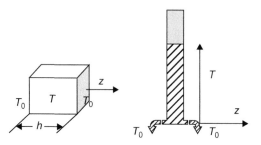

FIG. 3.25 Model for heat transfer from surface.

Equation (3.1.36) means the upper limit of the time increment Δt is inversely proportional to the thermal diffusivity $(\lambda/c\rho)$ and proportional to the square of the size of the cell $(h/2)$.

The same restriction on the time increment also comes from the heat transfer. This can be explained using the simple problem shown in Fig. 3.25. In this case, the whole thickness of the plate is modeled as one cell. Then, the heat flowing out from the surface to the atmosphere is given by

$$\Delta tq = 2\Delta t\beta s(T - T_0). \tag{3.1.37}$$

The temperature change produced by this heat flow ΔT becomes

$$\Delta T = -\Delta tq/(c\rho sh) = -2\Delta t\beta(T - T_0)/(c\rho h). \tag{3.1.38}$$

Noting that the temperature of the steel plate should not be lower than the ambient temperature T_0 and heat must flow from the steel to the atmosphere, $T + \Delta T > T_0$ must hold, that is,

$$T + \Delta T - T_0 = (T - T_0)(1 - 2\Delta t\beta/(c\rho h) > 0, \tag{3.1.39}$$

or,

$$\Delta t < (c\rho/\beta)(h/2). \tag{3.1.40}$$

Therefore, the upper limit of the time increment Δt is proportional to $c\rho/\beta$ and thickness h.

As discussed, the upper limit for the size of the time increment is determined by both heat conduction and heat transfer. It may be useful to compare these two limits in a welding problem, assuming that the material is mild steel and the values given in Table 3.1 are used for the heat conductivity λ, the specific heat c, and the density ρ. The heat transfer coefficient β is assumed to be 100 J/(sec m²°C), and the size of the cell corresponding to the size of the finite element, which will be discussed in Chapter 4, is changed from 10 mm to 1 mm. The values of $(c\rho/\lambda)(h/2)^2$ and $(c\rho/\beta)(h/2)$ are given for these cases. As seen from the table, the upper limit is given by the heat conduction, unless the heat transfer coefficient is extremely large. In this case, the time increment is

TABLE 3.3 Upper Limit of Time Increment Determined by Heat Conduction and Transfer

	$h=10$ mm	$h=4$ mm	$h=2$ mm	$h=1$ mm
Heat conduction $(c\rho/\lambda)h^2$	7.2 sec	1.15 sec	0.288 sec	0.072 sec
Heat transfer $(c\rho/\beta)h$	362 sec	145 sec	72.3 sec	36.2 sec

proportional to h^2; if $h=1$ mm, the time increment becomes about 0.072 sec, which is very small.

3.1.7 Summary

Basic knowledge of heat conduction, heat transfer, change of material properties with temperature, representative temperatures, and lengths is useful for understanding the theoretical background of the software for thermal analysis and stress analysis provided on the companion website. The discussions in this chapter will be helpful in Chapters 5 and 6 when preparing the input data for the FEM and when interpreting the validity of the computed results.

3.2 BASIC CONCEPTS OF MECHANICAL PROBLEMS IN WELDING

Generally, deformation and stress in structures such as bridges, buildings, etc. are produced by various kinds of applied loads and gravitational force. Therefore, deformation and stress of the structure under the action of these forces are estimated to assure safety.

On the other hand, when welding is applied to structures during their assembly, deformation and stress are induced in the structure by the expansion and shrinkage due to temperature changes caused by the heat introduced by the welding. During the thermal cycle induced by welding, mechanical properties of the material, such as Young's modulus, yield stress, etc., change with temperature. These changes are also causes of welding deformation and residual stress. In stress relief annealing, the structure is kept at a high temperature for a long period of time, and this process also produces the deformation and residual stress due to creep phenomena of the material.

As mentioned earlier, the most influential factors are thermal expansion and shrinkage, temperature-dependent material properties, and plastic and creep deformations. In this section, how to take these factors into consideration in various simulations [5, 17–19] by the FEM is briefly described using a simple three-bar model as an example, parallel with the basic concepts of mechanics.

3.2.1 Classification of Problems According to Dimensions

In this section, important ideas necessary to simulate the mechanical problems in welding are explained using a simple one-dimensional bar model as in Chapter 1. The structures to be simulated in practical engineering cases are welded components and structures that have complex three-dimensional geometry. In general, the required time for simulation increases as the dimension increases. Effort has been exerted to decrease computation time by idealizing three-dimensional problems into two-dimensional problems, such as plane stress, plane strain, plane deformation, or axis-symmetric problems. Typical examples that can be analyzed as one-, two-, and three-dimensional problems are presented in Figs. 3.26–3.32.

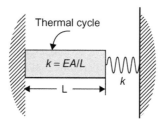

FIG. 3.26 Restrained elastic-plastic bar under thermal cycle (one-dimensional problem).

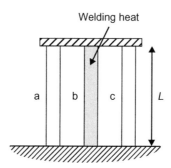

FIG. 3.27 Three-bar model under thermal cycle (one-dimensional problem).

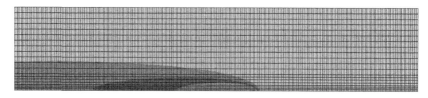

FIG. 3.28 Butt welding of thin plate (plane stress problem).

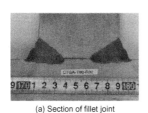

(a) Section of fillet joint

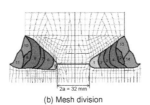

2a = 32 mm

(b) Mesh division

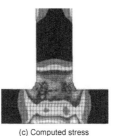

(c) Computed stress

FIG. 3.29   Fillet welding of thick plate (plane strain problem).

The examples shown in Chapters 1 and 2 are one-dimensional problems. The elastic-pastic restraint bar under the thermal cycle shown in Fig. 3.26 and the three-bar model subjected to thermal cycles shown in Fig. 3.27 are one-dimensional problems. Although these models are useful for understanding the basic phenomena in welding, they cannot be applied to quantitative prediction of distortion and residual stress in actual welded structures.

On the other hand, the following cases can be approximately handled as two-dimensional problems. One is the butt welding [4] of two thin plates in which the stress in the thickness direction can be negligible, as in the example shown in Fig. 3.28, which will be discussed in Chapter 6 as an example of welding simulation. Another example is a fillet weld of a thick plate [20] in which the strain induced in the welding direction is very small, as shown in Fig. 3.29. Similarly, the butt joints of pipes made by circumferential multipass welding [21] as shown in Fig. 3.30 or spot weldings can be regarded as axis-symmetric two-dimensional problems. In the butt joints of pipes, the change of welding condition in the circumferential direction is so small that the distribution of the stress and the deformation can be almost the same throughout the circumferential direction. Thus, they can be treated as functions of only the radial and axial coordinates.

However, when a pipe is joined to a plate slantingly, as shown in Fig. 3.31, the problem is not axis-symmetric, so it must be treated as a three-dimensional problem [22]. If the structures are assembled by thin plates and have a three-dimensional complex geometry, as shown in Fig. 3.32, they can be modeled using shell elements. In such cases, the analysis may be simplified using a shell model in which the stress is two-dimensional (plane stress), while the deformation is three-dimensional [23].

3.2.2 Variables and Equations Used to Describe Mechanical Problems

Prior to explaining the variables used to describe the mechanical phenomena in welding, general methods of structural or stress analysis without the influence

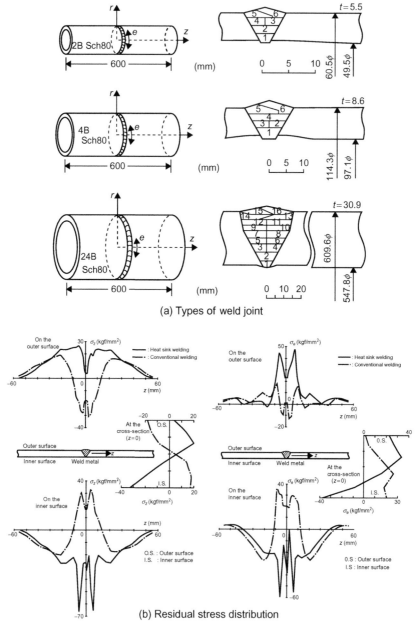

(a) Types of weld joint

(b) Residual stress distribution

FIG. 3.30 Butt welding of pipes (axis-symmetric problem).

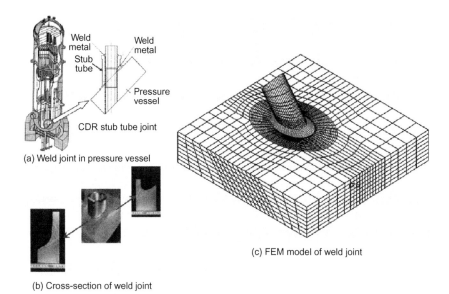

FIG. 3.31   Pipe joints in pressure vessel (three-dimensional problem).

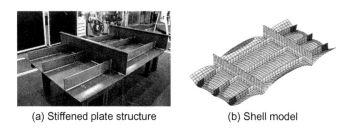

(a) Stiffened plate structure   (b) Shell model

FIG. 3.32   Stiffened plate structure (three-dimensional shell problem).

of temperature will be explained. The mechanical behavior of structures and machine parts can be described by the following three variables:

- Displacement: u
- Strain: ε
- Stress: σ

These variables should satisfy the following conditions:

- Strain-displacement relation: equation describing the relation between strain and displacement.
- Stress-strain relation: equation describing the relation between stress and strain.
- Equilibrium equation: the stress must satisfy this equation to keep equilibrium.

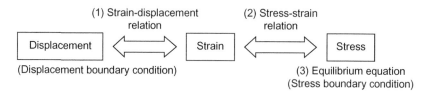

FIG. 3.33 Variables and relating equations in stress analysis.

There are three variables and three necessary conditions to be satisfied (see Fig. 3.33). At first glance, it may seem as if these are enough to solve the problem. It should be noted that both the displacement strain relation and the equilibrium condition should be satisfied not only inside but also on the surface or the boundary of the structure. The conditions required on the surface or the boundary are called boundary conditions, and there are two types, namely displacement boundary condition and stress boundary condition:

1. In a general stress analysis, there are three unknown variables, namely, the displacement, the strain, and the stress.
2. By solving the three governing equations, namely, the strain-displacement relation, the stress-strain relation, and the equilibrium equation, under the given boundary conditions, the displacement, the strain, and the stress are obtained.

Regardless of if the problem is a one-, two-, or three-dimensional problem, the variables and the equations to be solved are the same. The only difference is the number of coordinate axes and the components of the variables, such as the displacement, the strain, and the stress. These increase according to the dimension of the problem, as shown in Table 3.4.

3.2.3 Deformation and Stress in the Three-Bar Model

In the preceding section, it was shown that three equations—the strain-displacement relation, the stress-strain relation, and the equilibrium equation—must be solved under appropriate boundary conditions to obtain the deformation and stress produced in the structure. In this section, these equations are explained using the simple three-bar model given in Chapter 1. In welding, both the elastic deformation and the plastic deformation occur during the thermal cycle. For simplicity, only elastic deformation is considered first. The three-bar model shown in Fig. 3.34 is used as an example to show how to calculate deformation and stress. It is assumed that the force F is applied at its top and the center bar **b** is heated up by T °C. The three bars have the same length L, cross-sectional area A, and Young's modulus E. The bottom of the three bars are fixed, and their tops are joined together and move freely without rotation. The displacement at the top, and the stresses in the three bars, **a**, **b**, and **c**, are denoted u, σ_a, σ_b, and σ_c, respectively.

TABLE 3.4 Components of Variables in One-, Two-, and Three-Dimensional Problems

	Coordinate	Displacement	Strain	Stress
One-dimensional problem	x	$u(x)$	$\varepsilon(x)$	$\sigma(x)$
Plane-stress problem	y, v / x, u	$u(x, y),$ $v(x, y)$	$\varepsilon_x(x, y),$ $\varepsilon_y(x, y),$ $\gamma_{xy}(x, y),$ $\varepsilon_z(x, y)$	$\sigma_x(x, y),$ $\sigma_y(x, y),$ $\tau_{xy}(x, y),$ $\sigma_z(x, y) = 0$
Plane-strain problem	y, v / x, u	$u(x, y),$ $v(x, y)$	$\varepsilon_x(x, y),$ $\varepsilon_y(x, y),$ $\gamma_{xy}(x, y),$ $\varepsilon_z(x, y) = 0$	$\sigma_x(x, y),$ $\sigma_y(x, y),$ $\tau_{xy}(x, y),$ $\sigma_z(x, y)$
Axi-symmetric problem	z, w / θ / r, u	$u(r, z),$ $w(r, z)$	$\varepsilon_r(r, z),$ $\varepsilon_z(r, z),$ $\gamma_{rz}(r, z),$ $\varepsilon_\theta(r, z)$	$\sigma_r(r, z),$ $\sigma_z(r, z),$ $\tau_{rz}(r, z),$ $\sigma_\theta(r, z)$
Three-dimensional problem	z, w / y, v / x, u	$u(x, y, z),$ $v(x, y, z),$ $w(x, y, z)$	$\varepsilon_x(x, y, z),$ $\varepsilon_y(x, y, z),$ $\varepsilon_z(x, y, z),$ $\gamma_{xy}(x, y, z),$ $\gamma_{yz}(x, y, z),$ $\gamma_{zx}(x, y, z)$	$\sigma_x(x, y, z),$ $\sigma_y(x, y, z),$ $\sigma_z(x, y, z),$ $\tau_{xy}(x, y, z),$ $\tau_{yz}(x, y, z),$ $\tau_{zx}(x, y, z)$

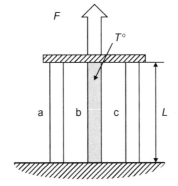

FIG. 3.34 Three-bar model subjected to external load and thermal load.

3.2.3.1 Strain-Displacement Relation

The strain in the three bars ε_a, ε_b, and ε_c can be related to the displacement u through the following equation.

$$\varepsilon_a = \varepsilon_b = \varepsilon_c = u/L, \tag{3.2.1}$$

where ε_a, ε_b, and ε_c are the strains that can be directly related to the displacement u and are called total strain or apparent strain. Since the total strain is produced as a result of the thermal expansion, elastic deformation, and plastic deformation, the total strain ε can be decomposed into the sum of the thermal strain ε^T, the elastic strain ε^e, and the plastic strain ε^p:

$$\varepsilon = \varepsilon^T + \varepsilon^e + \varepsilon^p. \tag{3.2.2}$$

Noting that plastic deformation is not considered at this stage, the strains in bars **a** and **c**, which are not heated, consist only of the elastic strain. The strain ε_b in bar **b**, which is heated, consists of the elastic strain ε_b^e and the thermal strain ε_b^T:

$$\varepsilon_b = \varepsilon_b^e + \varepsilon_b^T. \tag{3.2.3}$$

3.2.3.2 Stress-Strain Relation

When the thermal expansion coefficient of bar **b** is α, the elastic strain in bar **b** ε_b^e is written as

$$\varepsilon_b^e = \varepsilon_b - \alpha T. \tag{3.2.4}$$

Noting that stress is given by multiplying the elastic strain with the Young's modulus E, the stress in bars a, b and c are given as

$$\sigma_a = \sigma_c = E\varepsilon_a = E\varepsilon_c \tag{3.2.5}$$

$$\sigma_b = E(\varepsilon_b - \alpha_T). \tag{3.2.6}$$

3.2.3.3 Equilibrium Equation

From the condition that the force applied at the top F must balance with the stresses in the three bars, the following equilibrium equation is derived.

$$A(\sigma_a + \sigma_b + \sigma_c) = F. \tag{3.2.7}$$

By eliminating the strains and stresses from the previous three equations, the equation involves displacement u only can be derived as:

$$A(3Eu/L - E\alpha T) = F. \tag{3.2.8}$$

By rewriting,

$$Ku = F^*,\qquad(3.2.9)$$

where K is the stiffness and F^* is the force,

$$K = 3AE/L\qquad(3.2.10)$$

$$F^* = F + AE\alpha T.\qquad(3.2.11)$$

The force F^* is the sum of the external force F and the force due to the thermal expansion $AE\alpha T$.

The displacement u is obtained by solving Eq. (3.2.9):

$$u = (L/3AE)(F + AE\alpha T).\qquad(3.2.12)$$

Further, the strains and stresses in the bars can be calculated using the strain-displacement relation (Eq. 3.2.1), stress-strain relation (Eq. 3.2.5), and Eq. (3.2.6):

$$\varepsilon_a = \varepsilon_b = \varepsilon_c = (F + AE\alpha T)/(3AE)\qquad(3.2.13)$$

$$\sigma_a = \sigma_c = 1/3(F/A + E\alpha T)\qquad(3.2.14)$$

$$\sigma_b = 1/3(F/A - 2E\alpha T).\qquad(3.2.15)$$

3.2.4 Stress-Strain Relation in Welding

3.2.4.1 Strains to Be Considered in a Welding Problem

As discussed in the preceding section, the strain-displacement relation, the stress-strain relation, and the equilibrium equations must be solved under appropriate boundary conditions to obtain welding distortion and residual stress. Among these equations, only the stress-strain relation is different from ordinary stress analysis. The details of the stress-strain relation in a welding problem will be explained in the following.

The stress-strain relation is the equation describing how the strain is related to the stress. In welding, in which the material experiences high temperature during the thermal cycle, the strain is produced through different mechanisms, namely:

1. Elastic strain produced by stress that is reversible.
2. Thermal strain produced by temperature change.
3. Plastic strain produced by movement of dislocations that is irreversible.
4. Creep strain that appears when material is kept at high temperature under stress.

The key to derive the stress-strain relation is relating these strains produced through different processes to the stress.

The most important assumption is that the total strain ε can be decomposed into the sum of the thermal strain ε^T, the elastic strain ε^e, the plastic strain ε^p, and the creep strain ε^c, that is,

$$\varepsilon = \varepsilon^T + \varepsilon^e + \varepsilon^p + \varepsilon^c. \tag{3.2.16}$$

The plastic and the creep deformation are irreversible nonlinear problems. Even the elastic problem becomes a nonlinear problem when the Young's modulus and the thermal expansion coefficient change with temperature. Therefore, the welding problem is a highly nonlinear problem, and it is solved incrementally. In the incremental solution process, the time or the change of temperature is divided into small increments. The change of deformation and stress is computed in small increments and this process is repeated successively with the progress of time. Thus, the stress-strain relation must be written in incremental form using the strain and stress increments. The increments of the thermal, elastic, plastic, and creep strains will be briefly explained in the following.

Decomposition of Strain Increment

As shown in Eq. (3.2.16), the strain increment $\Delta \varepsilon$ can be decomposed into the thermal strain increment $\Delta \varepsilon^T$ and those of the elastic strain $\Delta \varepsilon^e$, plastic strain $\Delta \varepsilon^p$ and creep strain $\Delta \varepsilon^c$:

$$\Delta \varepsilon = \Delta \varepsilon^T + \Delta \varepsilon^e + \Delta \varepsilon^p + \Delta \varepsilon^c. \tag{3.2.17}$$

Further, the individual components of the strain increment can be derived from the fundamental equations as follows.

Thermal Deformation and Thermal Strain Increments

When a metal bar is heated from $0°C$ to $T°C$, its length becomes long by thermal expansion. The thermal expansion described as the strain is the thermal strain ε^T, and it is given as

$$\varepsilon^T = \alpha_0 T, \tag{3.2.18}$$

where α_0 is the average thermal expansion coefficient between the reference temperature, $0°C$ for example, and the temperature $T°C$, as shown in Fig. 3.35. According to Eq. (3.2.18), the thermal strain becomes zero when the temperature returns to the initial temperature. If the average thermal expansion coefficient α_0 is a function of the temperature, the thermal strain increment $\Delta \varepsilon^T$ due to a small temperature increment ΔT is given by

$$\Delta \varepsilon^T = \Delta \alpha_0 T + \alpha_0 \Delta T = \frac{d\alpha_0}{dT} \Delta T T + \alpha_0 \Delta T = \left(\frac{d\alpha_0}{dT} T + \alpha_0 \right) \Delta T \tag{3.2.19}$$

or

$$\Delta \varepsilon^T = \alpha \, \Delta T, \tag{3.2.20}$$

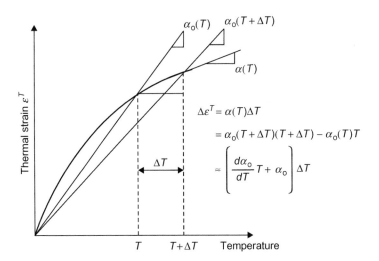

FIG. 3.35 Average and instantaneous thermal expansion coefficients.

where α is the instantaneous thermal expansion coefficient, defined as

$$\alpha = \frac{d\alpha_0}{dT} T + \alpha_0. \tag{3.2.21}$$

As explained here, the thermal expansion coefficient can be defined in two ways. In this text and the FEM program on the companion website, the instantaneous thermal expansion coefficient is used. Though the instantaneous thermal expansion coefficient is used in most commercial FEM code, it is recommended to be sure of the definition used before preparing the input data.

Elastic Deformation and Elastic Strain Increments

When the material is elastic, it follows Hook's law and the stress σ changes proportionally with the elastic strain ε^e:

$$\varepsilon^e = \sigma/E \quad \text{or} \quad \sigma = E\varepsilon, \tag{3.2.22}$$

where σ is the stress. As shown in Fig. 3.36, the elastic strain becomes zero when the stress is removed. If the Young's modulus E changes with the temperature T, the relation between the strain increment $\Delta\varepsilon^e$ and the stress increment $\Delta\sigma$ can be written in the same form as in the case of the thermal strain:

$$\Delta E \varepsilon^e + E \Delta \varepsilon^e = \frac{dE}{dT} \Delta T \varepsilon^e + E \Delta \varepsilon^e = \Delta\sigma. \tag{3.2.23}$$

Plastic Deformation and Plastic Strain Increments

When the material of the bar is elastic-plastic, the strain of the bar after the application of the tensile stress and its release show the cycle, as shown in

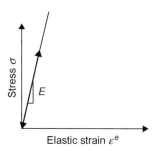

FIG. 3.36   Relation between elastic strain and stress.

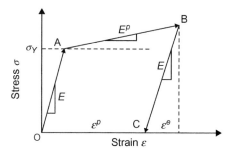

FIG. 3.37   Stress-strain curve in plastic deformation.

Fig. 3.37. When the applied stress is smaller than the yield stress σ_Y, the bar remains elastic as shown by line OA in Figure 3.37. The slope of the line OA is equal to the Young's modulus E. This slope becomes small when the stress becomes larger than the yield stress σ_Y, and it is called the tangent modulus and denoted E^P. Unloading process BC after the stress reaches its maximum value at point B is elastic and the slope of line BC is E. The strain recovered by the unloading is the elastic strain ε^e and the remaining strain is plastic strain ε^P. The stress-strain curve after the yielding is nonlinear in general and is approximated by a piecewise linear curve. Since the tangent modulus E^P, which is the ratio between the strain increment $\Delta\varepsilon$ and the stress increment $\Delta\sigma$, it is important and it will be explained in more detail.

For the time being, let's consider the elastic-plastic problem without the influence of temperature. Then the strain increment $\Delta\varepsilon$ becomes the sum of the elastic strain increment $\Delta\varepsilon^e$ and the plastic strain increment $\Delta\varepsilon^P$, that is,

$$\Delta\varepsilon = \Delta\varepsilon^e + \Delta\varepsilon^P. \tag{3.2.24}$$

Assuming that the Young's modulus is not a function of temperature or $dE/dT = 0$ for simplicity in Eq. (3.2.23), the relation between the elastic strain increment $\Delta\varepsilon^e$ and the stress increment $\Delta\sigma$ is given as

$$\Delta\varepsilon^e = \Delta\sigma/E. \tag{3.2.25}$$

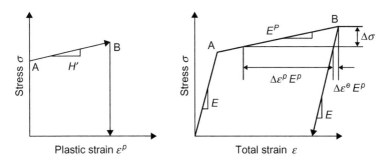

FIG. 3.38 Strain-hardening curve and stress increment under plastic loading and elastic unloading.

On the other hand, the plastic strain increment $\Delta\varepsilon^p$ can be derived from the yielding of the material and strain hardening. If the stress state is uniaxial as in the case of the bars, the material yields when the absolute value of the stress $|\sigma|$ reaches the yield stress σ_Y:

$$|\sigma| = \sigma_Y(\varepsilon^p). \tag{3.2.26}$$

If the deformation continues in the same direction after yielding, the stress σ increases with the deformation. This increase is caused by the increase of the plastic strain ε^p, and it is called the strain hardening. As shown in Fig. 3.38, the ratio H' between the increase of yield stress $\Delta\sigma_Y$ and that of the plastic strain $\Delta\varepsilon^p$ is called the strain-hardening coefficient and is defined by

$$\frac{d\sigma_Y}{d\varepsilon^p} = H', \tag{3.2.27}$$

and in the incremental form,

$$\Delta\sigma = H'\Delta\varepsilon^p. \tag{3.2.28}$$

By substituting Eqs. (3.2.25) and (3.2.28) into Eq. (3.2.24), the relation between the strain increment $\Delta\varepsilon$ and the stress increment $\Delta\sigma$ is obtained as

$$\Delta\varepsilon = \left(\frac{1}{E} + \frac{1}{H'}\right)\Delta\sigma = \frac{1}{E^p}\Delta\sigma \quad \text{or} \quad \Delta\sigma = E^p\Delta\varepsilon, \tag{3.2.29}$$

where the tangent modulus E^p is defined as

$$E^p = \frac{EH'}{E + H'}. \tag{3.2.30}$$

As discussed in Section 3.1, the yield stress of metals σ_Y is a function of both the plastic strain ε^p and temperature T, and the yield condition can be written as

$$\sigma = \sigma_Y(\varepsilon^p, T) \tag{3.2.31}$$

Rewriting Eq. (3.2.31) in the incremental form, the stress increment $\Delta\sigma$ is given in terms of the plastic strain increment $\Delta\varepsilon^p$ and the temperature increment ΔT:

$$\Delta\sigma = \frac{\partial\sigma_Y}{\partial\varepsilon^p}\Delta\varepsilon^p + \frac{\partial\sigma_Y}{\partial T}\Delta T = H'\Delta\varepsilon^p + H^T\Delta T, \tag{3.2.32}$$

where H^T is called the temperature-hardening coefficient and is defined as

$$H^T = \frac{\partial\sigma_Y}{\partial T}. \tag{3.2.33}$$

Creep Deformation and Creep Strain Increment

When a steel bar is kept at 600°C under constant tensile stress, it stretches with time like glass at a high temperature. Such deformation is called creep deformation. The relation between the creep strain ε^c and time t is illustrated in Fig. 3.39. The curve can be divided into three parts—namely, the first stage (transient creep), second stage (steady creep), and the third stage (accelerating creep). The creep in the postweld heat treatment is considered the transient creep since the stress is relaxed in a relatively short time. In general, the creep strain rate $\dot{\varepsilon}^c$ becomes large when the stress σ is large and the temperature T is high, and the creep strain is a function of the stress σ, temperature T, and time t. In postweld heat treatment in which both the stress and the temperature changes, the following equation proposed by Norton-Bailey is commonly used [9]:

$$\varepsilon^c = A\sigma^\lambda t^m. \tag{3.2.34}$$

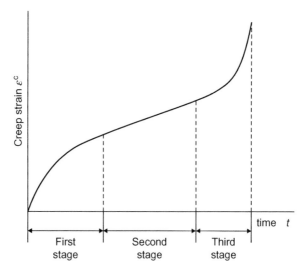

FIG. 3.39 Creep curve.

By differentiating with the time,

$$\dot{\varepsilon}^c = mA\sigma^\lambda t^{(m-1)}. \tag{3.2.35}$$

Further, by eliminating time t from Eqs. (3.2.34) and (3.2.35), the following relation can be derived:

$$\dot{\varepsilon}^c = mA^{1/m}\sigma^{\lambda/m}(\varepsilon^c)^{(1-1/m)}. \tag{3.2.36}$$

Since the incremental solution procedure is employed in the computation of the welding deformation and stress, the creep strain increment $\Delta\varepsilon^c$ is used. Assuming that the time increment Δt is small enough, the creep strain increment $\Delta\varepsilon^c$ can be approximated as the product of the time increment and the creep strain rate, that is,

$$\Delta\varepsilon^c \approx \dot{\varepsilon}^c \Delta t. \tag{3.2.37}$$

3.2.4.2 Incremental Stress-Strain Relation in Welding Problem

As discussed before, the strain increment can be decomposed into the thermal strain increment $\Delta\varepsilon^T$, the elastic strain increment $\Delta\varepsilon^e$, the plastic strain increment $\Delta\varepsilon^p$, and the creep strain increment $\Delta\varepsilon^c$. The individual components of the strain increment are given by Eqs. (3.2.20), (3.2.23), (3.2.32) and (3.1.37). By eliminating $\Delta\varepsilon^T$, $\Delta\varepsilon^e$, $\Delta\varepsilon^p$, and $\Delta\varepsilon^c$ from these equations, the incremental stress-strain relation is finally derived [5, 17, 19] as

$$\Delta\sigma = \frac{EH'}{E+H'}\Delta\varepsilon - \left\{\frac{EH'}{E+H'}\left(\alpha - \frac{1}{E^2}\frac{dE}{dT}\sigma\right) - \frac{E}{E+H'}\frac{d\sigma_Y}{dT}\right\}\Delta T - \frac{EH'}{E+H'}\dot{\varepsilon}^c \Delta t. \tag{3.2.38}$$

Though the detailed procedure to derive Eq. (3.2.28) is not given here, we know that the stress increment $\Delta\sigma$ consists of the following three parts:

1. Contribution from the strain increment $\Delta\varepsilon$:

$$D^{strain}\Delta\varepsilon = \frac{EH'}{E+H'}\Delta\varepsilon.$$

2. Contribution from the temperature increment ΔT through the thermal expansion and the temperature-dependent material properties:

$$D^{temp}\Delta T = -\left\{\frac{EH'}{E+H'}\left(\alpha - \frac{1}{E^2}\frac{dE}{dT}\sigma\right) - \frac{EH^T}{E+H'}\right\}\Delta T.$$

3. Contribution from the time increment Δt through creep:

$$D^{time}\Delta t = -\frac{EH'}{E+H'}\dot{\varepsilon}^c \Delta t.$$

This can be written symbolically as

$$\Delta\sigma = D^{strain}\Delta\varepsilon + D^{temp}\Delta T + D^{time}\Delta t. \tag{3.2.39}$$

3.2.5 Thermal Visco-Elasto-Plastic Problem in Three-Bar Model

To compute the stress and the deformation of structures, the strain-displacement relation, the stress-strain relation, and the equilibrium equation must be solved under appropriate boundary conditions. In this section, the procedure to solve the thermal visco-elasto-plastic problem in welding will be explained using the three-bar model as an example. As shown in Fig. 3.40, the force F is applied at the top in the upward direction, and the bar **b** in the center is heated by $T°C$. The three bars have the same length L, cross-sectional area A, and material properties. The bottoms of the three bars are fixed and their tops are joined together and move freely without rotation. The change of the deformation and stress during a small time increment Δt is considered incrementally. The force and the temperature change by ΔF and ΔT, respectively. The increments of the displacement at the top, the stresses in the three bars **a**, **b**, and **c** are denoted by Δu, $\Delta\sigma_a$, $\Delta\sigma_b$, and $\Delta\sigma_c$.

Strain-Displacement Relation:

$$\Delta\varepsilon_a = \Delta\varepsilon_b = \Delta\varepsilon_c = \Delta u/L. \tag{3.2.40}$$

Since bars **a** and **b** are not heated and remain elastic, the strain increments $\Delta\varepsilon_a$ and $\Delta\varepsilon_c$ are elastic strain. The strain increment $\Delta\varepsilon_b$ of bar **b**, which is heated, consists of the thermal, elastic, plastic, and creep strain increments. The stress

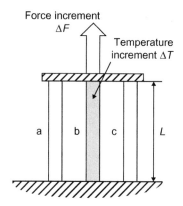

FIG. 3.40   Three-bar model subjected to external and thermal loads.

increment is given by Eq. (3.2.39). Noting these, stress increments in bars a, b and c are given by the following equations.

Stress-Strain Relation:

$$\Delta\sigma_a = \Delta\sigma_c = E\Delta\varepsilon_a = E\Delta\varepsilon_c \tag{3.2.41}$$

$$\Delta\sigma_b = D^{strain}\Delta\varepsilon_b + D^{temp}\Delta T + D^{time}\Delta t. \tag{3.2.42}$$

Considering that the force ΔF is acting on the top and that the stress increments in the three bars must balance with this force, the following equilibrium equation is derived:

Equilibrium equation:

$$A(\Delta\sigma_a + \Delta\sigma_b + \Delta\sigma_c) = \Delta F. \tag{3.2.43}$$

By eliminating the stress and the strain from Eqs. (3.2.39), (3.2.41), (3.2.42), and (3.2.43), the equilibrium equation is reduced to

$$A\{(2E + D^{strain})(\Delta u/L) + D^{temp}\Delta T + D^{time}\Delta t)\} = \Delta F. \tag{3.2.44}$$

Symbolically,

$$K\Delta u = \Delta F^*, \tag{3.2.45}$$

where K is the stiffness and ΔF^* is the load increment, and

$$K = (A/L)(2E + D^{strain}) \tag{3.2.46}$$

$$\Delta F^* = \Delta F + AD^{temp}\Delta T + AD^{time}\Delta t. \tag{3.2.47}$$

As seen from Eq. (3.2.47), the load increment ΔF^* involves contributions from the increment of external force ΔF, the temperature increment ΔT, and the time increment Δt.

3.2.5.1.1 Computation of Displacement, Strain, and Stress Increments:

By solving Eq. (3.2.45), the displacement increment Δu is obtained, that is,

$$\Delta u = L\{\Delta F/A + D^{temp}\Delta T + D^{time}\Delta t\}/(2E + D^{strain}). \tag{3.2.48}$$

Further, using the strain-displacement relation given by Eq. (3.2.40) and the stress-strain relation given by Eqs. (3.2.41) and (3.2.42), the strain increments and the stress increments are obtained:

$$\Delta\varepsilon_a = \Delta\varepsilon_b = \Delta\varepsilon_c = \Delta u/L \tag{3.2.49}$$

$$\Delta\sigma_a = \Delta\sigma_c = E(\Delta u/L) \tag{3.2.50}$$

$$\Delta\sigma_b = D^{strain}(\Delta u/L) + D^{temp}\Delta T + D^{time}\Delta t. \tag{3.2.51}$$

3.2.5.1.2 Computation of Total Displacement, Strain, and Stress

Further, the total displacement $u(t + \Delta t)$; the total strains $\varepsilon_a(t + \Delta t)$, $\varepsilon_b(t + \Delta t)$, $\varepsilon_c(t + \Delta t)$; and the total stresses $\sigma_a(t + \Delta t)$, $\sigma_b(t + \Delta t)$, $\sigma_c(t + \Delta t)$ at time $t + \Delta t$ can be obtained by adding the incremental values to those at time t, that is,

$$u(t + \Delta t) = u(t) + \Delta u \tag{3.2.52}$$

$$\varepsilon_a(t + \Delta t) = \varepsilon_a(t) + \Delta\varepsilon_a, \varepsilon_b(t + \Delta t) = \varepsilon_b(t) + \Delta\varepsilon_b, \varepsilon_c(t + \Delta t) = \varepsilon_c(t) + \Delta\varepsilon_c \tag{3.2.53}$$

$$\sigma_a(t + \Delta t) = \sigma_a(t) + \Delta\sigma_a, \sigma_b(t + \Delta t) = \sigma_b(t) + \Delta\sigma_b, \sigma_c(t + \Delta t) = \sigma_c(t) + \Delta\sigma_c. \tag{3.2.54}$$

3.2.6 Closing Remarks

In this section, fundamental relations in stress analysis, such as the strain-displacement relation, stress-strain relation, and equilibrium equation, were explained using a simple elastic three-bar model. The stress-strain relation in the welding problem was also discussed based on the fact that the strain is decomposed into thermal, elastic, plastic, and creep strains. Finally, the procedure for solving the thermal visco-elasto-plastic problem was demonstrated using the three-bar model as an example.

The fundamental relations shown for the one-dimensional problem can be extended to two- or three-dimensional problems, as summarized in Table 3.5.

TABLE 3.5 Correspondence in Stress-Strain Relation Between One Dimension and Two and Three Dimensions

Two and Three Dimensions	One Dimension
$[D^e]$	E
$\phi = \left\{ \dfrac{\partial f}{\partial \sigma} \right\}$	1
$S = [\phi][D^e]\{\phi\} + H'$	$E + H'$
$[D^{strain}] = [D^e] - \dfrac{[D^e]\{\phi\}[\phi][D^e]}{S}$	$D^{strain} = E - \dfrac{E^2}{E + H'} = \dfrac{EH'}{E + H'}$
$\{D^{temp}\} = [D^p]\left(\{\alpha\} - \dfrac{1}{E}\dfrac{dE}{dT}[D^e]^{-1}\{\sigma\}\right) - \dfrac{[D^e]\{\phi\}}{S}H^T$	$D^{temp} = \dfrac{EH'}{E + H'}(\alpha - \dfrac{1}{E^2}\dfrac{dE}{dT}\sigma) - \dfrac{EH^T}{E + H'}$
$\{D^{time}\} = -[D^{strain}]\{\varepsilon^c\}$	$D^{time} = -\dfrac{EH'}{E + H'}\varepsilon^c = -D^{strain}\varepsilon^c$
$\{\Delta\sigma\} = [D^{strain}]\{\Delta\varepsilon\} + \{D^{temp}\}\Delta T + \{D^{time}\}\Delta t$	$\Delta\sigma = D^{strain}\Delta\varepsilon + D^{temp}\Delta T + D^{time}\Delta t$

Those readers who wish to study the theoretical background may use this text as a guide along with the appropriate reference books [5, 9] or Appendix C, on the companion website.

REFERENCES

[1] Naka T. Welding, shrinkage and crack. Tokyo: Komine-Kogyo Publishers (in Japanese); 1950.

[2] Kihara H, Masubuchi K. Welding distortion and residual stress. Tokyo: Sanpo Publications (in Japanese); 1955.

[3] Watanabe M, Satoh K. Welding mechanics and its application. Tokyo: Asakura Publishing (in Japanese); 1965.

[4] Satoh K, Mukai Y, Toyoda M. Welding engineering. Tokyo: Rikogakusha Publishing (in Japanese); 1979.

[5] Satoh K, Ueda Y, Fujimoto T. Welding deformation and residual stress (Welding Series-3). Tokyo: Sanpo Publications (in Japanese); 1979.

[6] Japan Welding Society. Basic welding engineering. Tokyo: Maruzen Publishing (in Japanese); 1982.

[7] Fujino T. Heat conduction and thermal stress, seminar on structural engineering using computer, Vol. 2-4-B. Tokyo: Baifukan (in Japanese); 1972.

[8] Yagawa G. Introduction to FEM for flow and heat conduction analyses, series on basic and application of FEM, Vol. 8. Tokyo: Baifukan (in Japanese); 1983.

[9] Yagawa G, Miyazaki N. FEM for thermal stress, creep and thermal conduction analysis. Tokyo: Saiensu-sha (in Japanese); 1985.

[10] Ueda Y, Nakacho K, Kim Y, Murakawa H. Fundamentals of thermal-elastic-plastic-creep analysis and measurement of welding residual stress for numerical analysis. In Thermal Conduction Analysis Jl JWS 1986;55(6):336–48 (in Japanese).

[11] Japan Society of Thermophysical Properties. Handbook of thermophysical properties. Tokyo: Yokendo (in Japanese); 1990.

[12] The Japan Society of Mechanical Engineers. Elastic properties of metals (Technical Document). Tokyo: The Japan Society of Mechanical Engineers (in Japanese); 1980.

[13] Hara T, et al. The measurement of modulus of elasticity at high temperature and coefficient of thermal expansion for the comparison of the rate of frequency of thermal stress crackings in various steels: Study of thermal stress cracks in steel ingots III. ISIJ International 1963;49 (13):1885–91 (in Japanese).

[14] The Iron and Steel Institute of Japan (Creep Commission). Data collection of high temperature strength of metals. Tokyo: The Iron and Steel Institute of Japan (in Japanese); 1977.

[15] Garofalo F. ASTM SP 1952;129:10.

[16] Suzuki H. Up-to-date welding handbook. Tokyo: Sankaido Publishing (in Japanese); 1972.

[17] Ueda Y, Yamakawa T. Analysis of thermal elastic-plastic stress and strain during welding by FEM. Trans JWS 1971;2(2):90–100.

[18] Satoh K, Ueda Y, Fujimoto T. Welding deformation and residual stress (Welding Series-3). Tokyo: Sanpou Shuppan (in Japanese); 1979.

[19] Ueda Y, Nakacho K, Kim Y, Murakawa H. Fundamentals of thermal-elastic-plastic-creep analysis and measurement of welding residual stress for numerical analysis. In Thermal-Elastic-Plastic-Creep Analysis, Jl Japan Welding Society 1986b;55(7):399–410 (in Japanese).

[20] Nishikawa H, Serizawa H, Murakawa H. Actual application of FEM to analysis of large scale mechanical problem in welding. Sci Technol Welding Joining 2007;12(2):147–52.

[21] Ueda Y, Nakacho K, Shimizu T. Improvement of residual stresses of circumferential joint of pipe by heat-sink welding. Trans ASME Jl Pressure Vessel Technology 1986;108: 14–22.

[22] Nayama R, et al. Development of new measuring method of 3-dimensional residual stress and crack propagation simulation method for complex welding structure. Available from http://www.iae.or.jp/KOUBO/innovation/theme/pdf/h15-1-16.pdf.

[23] Murakawa H. Simulation of welding distortion. Tetsu-to-Hagane 2005;10(3):20–25 (in Japanese).

Chapter 4

The Finite Element Method

In this chapter, an outline of the finite element method (FEM), applied to the analysis of welding deformation and residual stress, is presented together with its history, important concepts, and its wide range of application. A flow of simulation is shown using simple example problems in welding mechanics. For validation of the computed results, theoretical solutions for fundamental problems are presented. Procedures to obtain rational results and to ensure successful simulation are also presented from identification of problems to preparation of data, execution of simulation, evaluation of results, extracting useful information and hints for solving the problem. Further, important checkpoints at each step of simulation are provided and troubleshooting procedures are presented.

4.1 FINITE ELEMENT METHOD AS A POWERFUL TOOL FOR A VARIETY OF PROBLEMS

The FEM was developed in the 1940s as a new computational method [1, 2, 3, 4, 5]. The method has been used as a powerful tool for stress analysis and a variety of scientific and engineering problems, such as heat conduction, electromagnetic problems, etc. Today, it is applied not only to linear problems but also to nonlinear problems, such as plasticity and ductile fracture, and it has become an indispensable tool for designing engineering products and evaluating their performance.

In this chapter, a simulation method for welding by the FEM is introduced. Welding simulation is a combination of heat conduction analysis and elastic-plastic stress analysis [6, 7, 8, 9, 10, 11]. Therefore, the computation consists of the following two parts:

1. Computation of temperature distribution, which changes with time (temperature analysis).
2. Computation of welding deformation, strain, and stress under the transient temperature field (stress analysis).

The advantage of the FEM is it represents objects as an assembly of small pieces of elements with very simple shapes, so the method can be applied to practical structures with complex geometry. The shapes of elements for one-, two-, and three-dimensional problems and examples of the mesh divisions are shown in Table 4.1. The bar element is used for one-dimensional problems, triangular and rectangular elements are for two-dimensional problems, and tetrahedron and hexahedron elements are used for three-dimensional problems. The point at the apex of each element is called the node. The temperature or displacement at the node is called the nodal temperature or nodal displacement. Stress and strain in the element are evaluated from the nodal displacements. Examples of mesh division are shown in Table 4.1. The welding of a fillet joint is idealized as a two- and three-dimensional

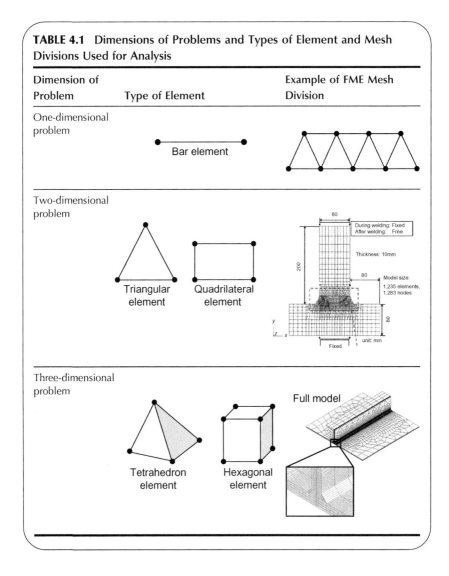

TABLE 4.1 Dimensions of Problems and Types of Element and Mesh Divisions Used for Analysis

Dimension of Problem	Type of Element	Example of FME Mesh Division
One-dimensional problem	Bar element	
Two-dimensional problem	Triangular element / Quadrilateral element	
Three-dimensional problem	Tetrahedron element / Hexagonal element	Full model

problem, and the models are divided using four-node quadrilateral elements and eight-node hexagonal solid elements, respectively.

The advantages of dividing the structure to be analyzed into small piece of elements are:

1. Complex structures of arbitrary geometry can be idealized by elements with simple shapes.
2. Since the element is geometrically simple, the mechanical behavior of the element can be expressed by simple equations.

These advantages allow us to describe the problem with simple equations. If the three-bar model and the three-cylinder model discussed in Chapter 3 are taken as examples, the equilibrium equation and the thermal balance equation can be described as the following form,

$$K \Delta u = \Delta F^* \tag{4.1.1}$$

$$[C]\left\{\frac{dT}{dt}\right\} + [K]\{T\} = \{Q\}, \tag{4.1.2}$$

where Δu and T are the increments of displacement and temperature and

$$\left\{\frac{dT}{dt}\right\} = \begin{Bmatrix} dT_A/dt \\ dT_B/dt \\ dT_C/dt \end{Bmatrix} \tag{4.1.3}$$

$$\{T\} = \begin{Bmatrix} T_A \\ T_B \\ T_C \end{Bmatrix}. \tag{4.1.4}$$

In the case of the three-bar model, the bars are connected to a rigid body at the top. Thus, the governing equation, Eq. (4.1.1), includes only one unknown value. In contrast to this, Eq. (4.1.2) for the three cylinders connected by tubes contains six unknowns $(dT_A/dt, dT_B/dt, dT_C/dt)$ and (T_A, T_B, T_C). The basic form of the governing equation for simulation is the same regardless of the size or the complexity of the welded structure.

4.2 TYPES OF PROBLEMS AND THE CORRESPONDING BASIC EQUATIONS

To understand the basic differences and similarities among the equations that describe different problems, a simple example problem consisting of a spring or a bar is used. Table 4.2 shows simple models that consist of a spring (spring constant K), a mass (mass M), and a dumper (dumping C). In (a), the model consists of only a single spring, and the basic equation has the same form as

TABLE 4.2 Basic Form of Equations for Various Types of Problems

Mechanical Components Composing Problem	Problems to be Applied	Equation
(a) spring 	Stress analysis Steady heat conduction problem Static electric field analysis	$Ku = f$
(b) spring + mass 	Vibration analysis	$M\dfrac{d^2u}{dt^2} + Ku = f$
(c) spring + dumper 	Transient heat conduction analysis	$C\dfrac{du}{dt} + Ku = f$
(d) spring + mass + dumper 	Vibration analysis	$M\dfrac{d^2u}{dt^2} + C\dfrac{du}{dt} + Ku = f$

Eq. (4.1.1) for the stress analysis. In (c), the model is composed of a spring and a dumper, and the basic equation includes a time derivative in the same form as Eq. (4.1.2), which describes the heat conduction. In (b) and (d), mass is added to the model, and the second derivative is included in the equation. These are the basic equations for vibration analysis.

In welding simulation, which consists of heat conduction analysis and stress analysis, equations of type (a) and (c) in Table 4.2 must be solved. In the simulation of resistance spot welding, the heat is generated by Joule heating. Thus, the electric current must be analyzed in addition to the temperature field and the stress field. As seen from Table 4.2, the equation for the electric current analysis has the same form as for the stress analysis, and additional effort is not required for the simulation of the resistance spot welding.

4.3 BASIC CONCEPTS OF THE VARIATIONAL PRINCIPLE

The main part of stress analysis is to solve the following equation for the unknown displacement u:

$$Ku = f. \tag{4.3.1}$$

If the problem is very simple, such as the case of a spring shown in Table 4.2, the solution procedure is simple. However, when the structure to be solved is large and complex, it may not be easy to understand how Eq. (4.3.1) is derived. Equation (4.3.1) can be derived very systematically when the variational principle is employed. For readers who want to solve welding problems using commercial software, it may not be necessary to understand the variational principle in detail. It is enough to understand its outline. The variational principle is just another expression of Eq. (4.3.1) in the following form.

The solution u is the displacement that minimizes the following function Π:

$$\Pi(u) = \frac{1}{2}Ku^2 - fu. \tag{4.3.2}$$

We can readily show that the above statement is equivalent to Eq. (4.3.1). Function $\Pi(u)$ becomes the minimum value when the following conditions are satisfied:

$$\frac{d\Pi}{du} = 0 \tag{4.3.3}$$

$$\frac{d^2\Pi}{du^2} > 0. \tag{4.3.4}$$

From Eq. (4.3.3), it can be shown that

$$\frac{d\Pi}{du} = Ku - f = 0. \tag{4.3.5}$$

Eq. (4.3.5) is the same as Eq. (4.3.1). Since the second derivative of Π is given by the following equation

$$\frac{d^2\Pi}{du^2} = K,$$ (4.3.6)

and the spring constant K is a positive value, Eq. (4.3.4) is satisfied.

The important point here is the physical meaning of Eq. (4.3.2). The first term on the right side of the equation is the strain energy stored in the spring and the second term represents the potential energy of the external force f. Thus, Eq. (4.3.2) is the total potential energy of the mechanical system. Since the total potential energy becomes the minimum value when the displacement u is the solution, this principle is called the principle of minimum potential energy. Among various variational principles, the principle of virtual work is also useful especially for nonlinear problems such as in welding. Based on these two principles, most of the FEM is constructed.

4.4 HOW TO SOLVE A PROBLEM WITH MORE THAN ONE ELEMENT

As shown in Table 4.1, welding problems are two- or three-dimensional problems, and they are modeled with more than one element. The procedure to solve such problems is explained using two linearly connected bars as an example.

4.4.1 Equilibrium Equation of a Bar

As shown in Table 4.1, the simplest element is a bar element in a one-dimensional problem. The equilibrium equation for the single bar shown in Fig. 4.1 can be derived in the following manner. The length, the cross-sectional area, and the Young's modulus of the bar are denoted by L, A, and E, respectively. The nodes at both ends of the bar are labeled node 1 and node 2. The forces acting on these nodes (nodal forces) and the displacements at these nodes (nodal displacements) are denoted by f_1, f_2, u_1, and u_2. The axial force, the strain, and the stress in the bar are denoted by f, ε, and σ.

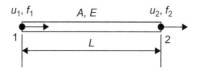

FIG. 4.1   Structure with one element.

The equilibrium equation of the bar is derived by the procedure explained in Section 3.2. To solve the problem, three governing equations—namely, the strain-displacement relation, the stress-strain relation, and the equilibrium— must be satisfied along with the appropriate boundary conditions.

- Strain-displacement relation:

$$\varepsilon = (u_2 - u_1)/L \tag{4.4.1}$$

- Stress-strain relation:

$$\sigma = f/A = E\varepsilon \tag{4.4.2}$$

- Equilibrium equation:

$$\text{node 1: } f_1 + f = 0, \quad \text{node 2: } f_2 - f = 0 \tag{4.4.3}$$

By substituting Eqs. (4.4.1) and (4.4.2) into Eq. (4.4.3), the strain and the stress are eliminated, and the equilibrium equations are expressed in terms of displacement only, that is,

$$(EA/L)(u_1 - u_2) = f_1$$
$$(EA/L)(u_2 - u_1) = f_2. \tag{4.4.4}$$

These equations may be rewritten in matrix form as

$$\begin{bmatrix} k & -k \\ -k & k \end{bmatrix} \begin{Bmatrix} u_1 \\ u_2 \end{Bmatrix} = \begin{Bmatrix} f_1 \\ f_2 \end{Bmatrix}. \tag{4.4.5}$$

Equation (4.4.5) represents the characteristics of the element, and the matrix

$$\begin{bmatrix} k & -k \\ -k & k \end{bmatrix}$$

is the stiffness matrix of the element, where $k = EA/L$.

4.4.2 Equilibrium Equations of Two Bars

Figure 4.2 shows that the structure consists of two linearly connected bars, a and b. The lengths of the bars are L_a and L_b. They have the same cross-sectional area A and Young's modulus E. The nodes are sequentially numbered from the left side. The nodal displacements and forces at each node are denoted

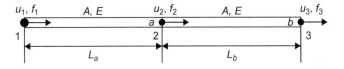

FIG. 4.2 Structure with two linearly connected bars.

respectively by $u_1, u_2, u_3, f_1, f_2,$ and f_3. The axial forces, the strains, and the stresses in each bar are denoted respectively by $f_a, f_b, \varepsilon_a, \varepsilon_b,$ and σ_a, σ_b. The three governing equations for this problem are as follows.

• Strain-displacement relation:

$$\varepsilon_a = (u_2 - u_1)/L_a$$
$$\varepsilon_b = (u_3 - u_2)/L_b \tag{4.4.6}$$

• Stress-strain relation:

$$\sigma_a = f_a/A = E\varepsilon_a$$
$$\sigma_b = f_b/A = E\varepsilon_b \tag{4.4.7}$$

• Equilibrium equation:

$$f_1 + f_a = 0$$
$$f_2 - f_a + f_b = 0 \tag{4.4.8}$$
$$f_3 - f_b = 0$$

By substituting Eqs. (4.4.6) and (4.4.7) into Eq. (4.4.8), the strain and stress are eliminated, and the equilibrium equations are expressed in terms of displacement only, that is,

$$(k_a u_1 - k_b u_2) = f_1$$
$$\{-k_a u_1 + (k_a + k_b)u_2 - k_b u_3\} = f_2 \tag{4.4.9}$$
$$(-k_b u_2 + k_b u_3) = f_3.$$

These equations may be rewritten in matrix form as

$$[K]\{u\} = \begin{bmatrix} k_a & -k_a & 0 \\ -k_a & k_a + k_b & -k_b \\ 0 & -k_b & k_b \end{bmatrix} \begin{Bmatrix} u_1 \\ u_2 \\ u_3 \end{Bmatrix} = \begin{Bmatrix} f_1 \\ f_2 \\ f_3 \end{Bmatrix}, \tag{4.4.10}$$

where $k_a = EA/L_a$ and $k_b = EA/L_b$.

Equation (4.4.10) is the equilibrium equation for two linearly connected bars. Comparing the stiffness matrices of the two cases in Eqs. (4.4.5) and (4.4.10), it is readily seen that the stiffness matrix of the two connected bars is composed by superposition of the stiffness matrices of bars **a** and **b**. This suggests that the equilibrium equation of a complex structure composed of many elements may be derived by straightforward superposition of the stiffness matrix of each element according to the same rule as shown by Eq. (4.4.10).

As explained, the equilibrium equation can be derived from the strain-displacement relation, the stress-strain relation, and the equilibrium equation. However, the boundary condition has not been considered. To understand the importance of the boundary condition, let's examine whether Eq. (4.4.10) can be solved for displacements u_1, u_2, and u_3. As Eq. (4.4.10) is a set of simultaneous equations, the determinant of the coefficient matrix should not vanish to get the solution. In this case, the determinant becomes zero as shown:

$$|K| = k_a(k_a + k_b)k_b - (k_a)^2 k_b - k_a(k_b)^2 = 0. \qquad (4.4.11)$$

This implies that the solution cannot be obtained.

Let's examine the case where the boundary conditions are considered. Since the left end of the bar is fixed and force P is acting at point 3, and the force is not acting at point 2, the boundary conditions can be given by the following equations:

$u_1 = 0$ at node 1 (geometrical boundary condition)
$f_2 = 0$ at node 2 (force boundary condition)
$f_3 = P$ at node 3 (force boundary condition)

Considering these boundary conditions, Eq. (4.4.10) is reduced to

$$\begin{bmatrix} k_a & -k_a & 0 \\ -k_a & k_a + k_b & -k_b \\ 0 & -k_b & k_b \end{bmatrix} \begin{Bmatrix} 0 \\ u_2 \\ u_3 \end{Bmatrix} = \begin{Bmatrix} f_1 \\ 0 \\ P \end{Bmatrix}. \qquad (4.4.12)$$

Noting that the displacements u_2 and u_3 are unknown and u_1 is specified, Eq. (4.4.12) is rearranged as

$$-k_a u_2 = f_1 \qquad (4.4.13)$$

$$\begin{bmatrix} k_a + k_b & -k_b \\ -k_b & k_b \end{bmatrix} \begin{Bmatrix} u_2 \\ u_3 \end{Bmatrix} = \begin{Bmatrix} 0 \\ P \end{Bmatrix}. \qquad (4.4.14)$$

In this case, the determinant of the coefficients matrix of Eq. (4.4.14) is not zero as shown:

$$|K| = (k_a + k_b)k_b - (k_b)^2 = k_a k_b > 0. \qquad (4.4.15)$$

Therefore, the equation can be solved for u_2, u_3:

$$u_2 = P/k_a, \quad u_3 = P(1/k_a + 1/k_b). \qquad (4.4.16)$$

Further, the unknown reaction force f_1 is obtained using Eq. (4.4.13), that is,

$$f_1 = -P. \qquad (4.4.17)$$

The strains and stresses in each bar can be obtained by substituting u_2, u_3 into Eqs. (4.4.6) and (4.4.7):

$$\begin{aligned} \varepsilon_a &= (u_2 - u_1)/L_a = P/AE \\ \varepsilon_b &= (u_3 - u_2)/L_b = P/AE \\ \sigma_a &= E\varepsilon_a = P/A \\ \sigma_b &= E\varepsilon_b = P/A. \end{aligned} \qquad (4.4.18)$$

So far, the outline of the procedure for stress analysis using the FEM has been presented. As most of the above procedure can be applied to heat conduction analysis, details for this analysis are not given here.

4.5 INCREMENTAL METHOD FOR NONLINEAR PROBLEMS

In welding problems, the temperature, displacement, strain, and stress in a structure change with time in a complex manner. To solve this type of problem, the incremental method is employed. Figure 4.3 schematically shows the incremental solution procedure. In this figure, Δt is the time increment and computation is started from time $t = 0$ and proceeds step by step with the time increment Δt. In each step, increments of temperature, displacement, strain, and stress ($\Delta T, \Delta u, \Delta \varepsilon, \Delta \sigma$ respectively) are computed. The values at time $t + \Delta t$ can be estimated by adding these increments to those at time t according to the following equations:

$$T(t + \Delta t) = T(t) + \Delta T \tag{4.5.1}$$

$$u(t + \Delta t) = u(t) + \Delta u \tag{4.5.2}$$

$$\varepsilon(t + \Delta t) = \varepsilon(t) + \Delta \varepsilon \tag{4.5.3}$$

$$\sigma(t + \Delta t) = \sigma(t) + \Delta \sigma. \tag{4.5.4}$$

Since there is some interaction between the temperature field and the stress field, temperature and stress must be solved simultaneously in a strict sense. Noting that the heat generated through the mechanical process is relatively small, the thermal and stress analyses can be performed individually as shown in Fig. 4.4. According to this procedure, first the temperature field is computed from the start of welding until the structure cools down to room temperature. The temperature at each time step is stored in a file in the computer. In stress analysis, the temperature is imported from the file to compute the deformation and stress at each time step. This process is continued from the start of the welding until the complete cool down.

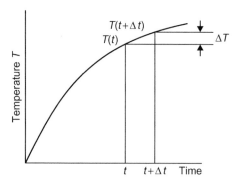

FIG. 4.3 Incremental method for nonlinear problems.

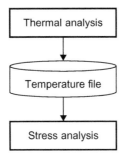

FIG. 4.4 Serial thermal and stress analysis in incremental method.

4.6 SIMPLE EXAMPLES OF ANALYZING THERMAL ELASTIC-PLASTIC-CREEP BEHAVIOR

4.6.1 Bar Fixed at Both Ends Under a Thermal Cycle

According to the procedure for solving the problem by the FEM explained in the preceding section, several examples related to welding are demonstrated. The first example is a metal bar fixed at both ends, as shown in Fig. 4.5, which is subjected to thermal cycles. A bar with a rectangular cross-section is set along the x-axis and fixed at both ends. Its dimensions are: length = 200 mm, width of the cross-section = 20 mm, and thickness = 10 mm. The bar is subjected to four different thermal cycles. As shown in Fig. 4.6, the bar is heated up from 0°C to 50°C, 150°C, 300°C, or 600°C, and cooled down to 0°C. The thermal expansion coefficient α and the Young's modulus E do not change with temperature, and they are assumed to be 1×10^{-5} 1/°C and 200 GPa, respectively. It is assumed that the material is elastic perfectly plastic material without strain hardening, and the yield stress changes with the temperature as shown in Fig. 4.7. It is kept constant at 200 MPa between 0°C and 300°C, then linearly decreases to 10 MPa from 300°C to 775°C, and again kept constant at 10 MPa above 775°C.

4.6.1.1 Mesh Division

Though the problem is a one-dimensional problem, it can be dealt with as a two-dimensional problem in the x–y plane, or as a three-dimensional problem. In this section, the bar is analyzed as a two-dimensional plain stress problem. The bar is modeled as a long rectangular plate and its thickness and width are 10 mm and 20 mm. This plate is divided by a square element of 5 mm × 5 mm. This implies 20 divisions in the x direction and 4 divisions in the y direction, and the total number of elements becomes 80. As computation time for analysis increases generally with the number of elements or the number of unknown displacements, computational time can be reduced if the plate is modeled using a small number of elements. Since the plate is symmetric in length and width, it is enough to analyze one-quarter of the

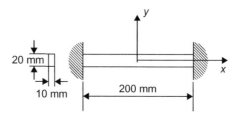

FIG. 4.5   Metal bar fixed at both ends subjected to thermal cycles.

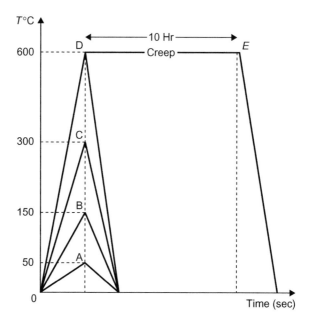

FIG. 4.6   Specified thermal cycles to the bar.

plate considering the symmetry. In this way, one-quarter of the plate is modeled using only 20 elements as shown in Fig. 4.8. In the figure, the elements and nodes are numbered sequentially from 1 to 20 and from 1 to 33, respectively.

4.6.1.2 Boundary Conditions

If the object to be analyzed is not constrained at all, the displacement cannot be determined because the simultaneous equation becomes singular. In three-dimensional problems, three translations and three rotations in the x, y, and z directions should be restrained to prevent rigid body motion. Similarly, for two-dimensional problems, two translations in the x and y directions and one rotation around the z-axis must be fixed. When the symmetry of the model is considered, the procedure to apply the constraint to prevent rigid body motion is as follows. First, give the constraints

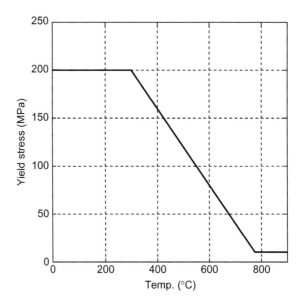

FIG. 4.7 Temperature-dependent yield stress of metal.

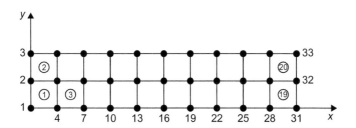

FIG. 4.8 Mesh division for one-quarter of the bar.

from the symmetry, then add the constraints for the fixed part of the model. After this, examine whether the rigid body motion is fixed. If not, add a sufficient number of constraints, that is:

1. Because of the symmetry with respect to the x-axis, the displacements in the y direction at nodes 1, 4, 7, ..., 31 are fixed. Similarly, because of the symmetry with respect to the y-axis, those in the x direction at nodes 1, 2, and 3 are fixed.

2. As the bar is fixed at both ends, the displacements in the x direction at nodes 31, 32, and 33 should be constrained.

3. Since the constraints imposed by (1) and (2) prevent the rigid body motion of the bar, no more constraint is necessary to control rigid body motion.

4.6.1.3 Thermal Cycles

In the example analysis of welding discussed in Chapter 5, first the heat conduction analysis is performed for temperature changes in the specimen due to the heat input provided by welding. The output is maintained as nodal temperatures in a file. In the following stress analysis, changes in nodal temperature at each elapse of time are supplied from the file and the analysis continues step by step with small time or temperature increment. Since the temperature distribution in the bar is assumed to be uniform in this example, it is not necessary to compute the temperature by heat conduction analysis. The file for the temperature data is easily prepared by hand. In the case where the bar is heated up from 0°C to 50°C and cooled down to 0°C, temperature data for 10 steps is required if the temperature increment in the computation is 10°C.

4.6.2 Thermal Elastic Behavior of a Bar Fixed at Both Ends

As discussed in Section 1.2.3, the residual stress of a bar is different according to the magnitude of the maximum temperature in the thermal cycle. When the maximum temperature is 50°C, the stress in the bar is below the elastic limit and the bar behaves elastically. Figure 4.9 shows the stress distribution along the elements from 1, 3, 5–19 when the temperature is 50°C. As expected before the FEM analysis, the stresses is uniform along its length:

$$\sigma_x = -E\alpha\Delta T = -100 \text{ MPa}.$$

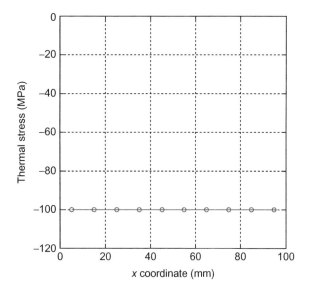

FIG. 4.9   Stress distribution in restrained bar at 50°C under thermal cycle.

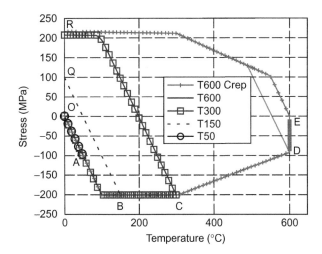

FIG. 4.10 Stress temperature curve of restrained bar under thermal cycle.

The change of the stresses in the thermal cycle is plotted in Fig. 4.10 as a straight line, O-A. Since the bar remains in an elastic state, the stress increases linearly with the temperature. It is also seen from the figure that the maximum stress is −100 MPa, below the yield stress of the bar, which is 200 MPa. In the cooling process, the stress decreases following the same line O-A as in the heating process, and the stress returns to zero when the temperature returns to 0°C.

4.6.3 Thermal Elastic-Plastic Behavior of a Bar Fixed at Both Ends

Using the same bar discussed in the previous section, the procedure of analyzing the elastic-plastic behavior of the bar and the result is shown. The same mesh division, boundary condition, and material properties as those in the previous section are used. For the elastic-plastic problem, the three different maximum temperatures in the thermal cycles are studied. In each case, the maximum temperature is assumed to be 150°C, 300°C, and 600°C. The results of the analysis are shown in Fig. 4.10. The lines go through O-A-B-Q, O-A-B-C-R, and O-A-B-C-D-R and indicate the relation between the stress σ_x and the temperature T in the three cases. At temperature 100°C in the heating process, the stress reaches −200 MPa, which is the yield stress of the material. When the temperature exceeds 300°C, the yield stress decreases, as shown in Fig. 4.7. The stress in the bar decreases in the same manner.

The behavior in the cooling process is studied here. When the maximum temperature is 150°C, the stress increases with cooling, and it becomes 100 MPa in tension after cooling down to 0°C. Since the stress is below the yield stress in this case, the bar behaves elastically in the cooling stage.

If the maximum temperature is 300°C, the stress becomes 200 MPa in tension when the bar is cooled down to 100°C. The cooling process continues to 0°C keeping the same stress that is equal to the yield stress. Finally, 200 MPa remains as the tensile residual stress.

4.6.4 Thermal Elastic-Plastic Creep Behavior of a Bar Fixed at Both Ends

Using the same bar as analyzed in the preceding sections, an analysis is performed under the condition that the bar is heated up to 600°C from room temperature and held at this temperature for 10 hours, as shown by line D-E in Fig. 4.6. For this analysis, the material property for creep is necessary. The equivalent creep strain rate $\dot{\bar{\varepsilon}}^c$ is assumed to be a function of the temperature and the equivalent stress $\bar{\sigma}$, as shown in Table 4.3. The temperature stress curve for this case is shown by line O-A-B-C-D-E-R in Fig. 4.10. It is seen from this figure that the stress approaches zero during the holding time. Figure 4.11 shows the relaxation of the stress after the start of holding. It is found that the stress decreases almost to zero after 2 hours of holding at 600°C. However, it should be noted that during the cooling process in which the bar is cooled down to room temperature, the shrinkage produces tensile stress and results in tensile yield stress since the bar is fixed at both ends.

Using a very simple example, the procedure of analyzing the elastic-plastic behavior under various thermal cycles and the evolution of the stress during the thermal cycle is examined in this section. Readers can analyze the same problem themselves using the program and the input data provided on this book's website. The instructions for how to perform the computation are given in Chapter 5. Examples for dealing with welding, including both thermal and stress analyses, are presented in Chapter 6. Other examples such as the simulation of the RRC test and the slit welding test and an

TABLE 4.3 Dependency of Creep Strain Rate on Temperature and Stress

Creep Strain Rate	Temperature $T°C$	Temperature-Dependent Term $F(T)$	Equivalent Stress $\bar{\sigma}$	Stress-Dependent Term $g(\bar{\sigma})$
$\dot{\bar{\varepsilon}}^c = A \cdot f(T) \cdot g(\bar{\sigma})$	0	0	0	0
	300	0	50	0.5
$\dot{\varepsilon}_i^c = \frac{3}{2} \frac{\sigma_i'}{\bar{\sigma}} \dot{\bar{\varepsilon}}^c$	600	1	100	1
$A = 3.6 \times 10^{-7} (1/s)$	1000	1	200	1

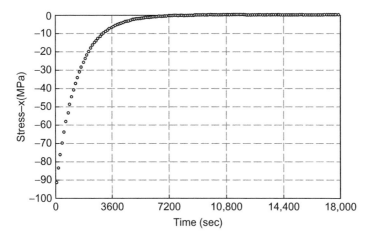

FIG. 4.11 Relaxation of stress during holding at 600°C.

application of inherent strain theory to a three-bar-model are also included on the companion website.

4.7 BASIC THEORETICAL SOLUTIONS TO VALIDATE RESULTS OBTAINED BY THE FEM

It is very important to evaluate whether the results of FEM analysis are reasonable and accurate enough to be accepted. For this purpose, the basic theoretical solutions for heat conduction and elastic-plastic mechanics are presented in this section.

4.7.1 Temperature Distribution Due to a Concentrated Heat Source

In heat conduction problems there are three typical heat sources: the instantaneous point source, the instantaneous line source, and the instantaneous plane source, as shown in Table 4.4. On the other hand, the heat sources in welding move with the torch, and they are different from the instantaneous heat sources in this sense. As schematically illustrated in Table 4.5, when the torch moves with a high speed on a thin plate, the temperature distribution far enough from the torch is similar to that produced by an instantaneous plane heat source. Thus, the problem may be treated as a one-dimensional problem. Noting this, temperature distributions produced by welding on a thin plate and a thick plate are calculated, and the relations between the distance from the torch and the maximum temperature attained are studied. Further, it can be shown that the inherent strain and the residual stress in the welding direction are estimated using these maximum temperatures.

TABLE 4.4 Transient Temperature Distribution Generated by Instantaneous Heat Source (Initial Temperature = 0 $T°C$)

	Instantaneous Point Heat Source	Instantaneous Line Heat Source	Instantaneous Plane Heat Source
Form of heat source $\begin{smallmatrix}z\\ \nearrow y\\ \llcorner\rightarrow x\end{smallmatrix}$			
Dimension of problem	Three-dimensional	Two-dimensional	One-dimensional
Transient temperature: T	$T(r,t) = \dfrac{q_3}{c\rho} \dfrac{e^{-\frac{r^2}{4kt}}}{(4\pi kt)^{3/2}}$	$T(r,t) = \dfrac{q_2}{c\rho} \dfrac{e^{-\frac{r^2}{4kt}}}{(4\pi kt)}$	$T(r,t) = \dfrac{q_1}{c\rho} \dfrac{e^{-\frac{r^2}{4kt}}}{(4\pi kt)^{1/2}}$
Maximum temperature: T_m	$T_m(r) = \dfrac{q_3}{c\rho}\left(\dfrac{3}{2\pi e}\right)^{3/2}\dfrac{1}{r^3}$	$T_m(r) = \dfrac{q_2}{c\rho}\left(\dfrac{1}{\pi e}\right)\dfrac{1}{r^2}$	$T_m(r) = \dfrac{q_1}{c\rho}\left(\dfrac{1}{2\pi e}\right)^{1/2}\dfrac{1}{r}$
Distance: r	$r = (x^2 + y^2 + z^2)^{1/2}$	$r = (x^2 + z^2)^{1/2}$	$r = x$
Thermal diffusivity: k	$k = \dfrac{\lambda}{c\rho}$	$k = \dfrac{\lambda}{c\rho}$	$k = \dfrac{\lambda}{c\rho}$

Unit for heat source: q_1 (J), q_2 (J/m), q_3 (J/m^2)

4.7.2 Temperature Distribution on a Butt-Welded Joint of a Thin Plate

In this section, the temperature distribution on a butt-welded thin plate on the x-y plane as shown in Fig. 4.12 is considered. If the welding speed is very high, the heat supplied from the moving torch can be idealized as an instantaneous heat source uniformly distributing along the weld line. In this case, the temperature distribution is uniform in the y direction and heat flows only in the x direction. Then the problem may be treated as a one-dimensional problem in which the heat source is represented by a plane heat source. In such cases, the transient temperature distribution is given by the following equation, if the initial temperature is assumed to be zero and the heat diffusion from the plate surface is neglected:

$$T(x,t) = \frac{q_1}{c\rho} \frac{e^{-\frac{x^2}{4kt}}}{(4\pi kt)^{1/2}}. \qquad (4.7.1)$$

TABLE 4.5 Characteristics of Transient Thermal Field in Welding and Dimension of Problem

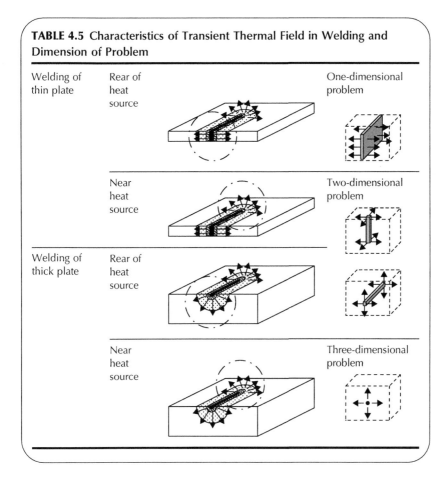

Welding of thin plate	Rear of heat source	One-dimensional problem
	Near heat source	Two-dimensional problem
Welding of thick plate	Rear of heat source	Two-dimensional problem
	Near heat source	Three-dimensional problem

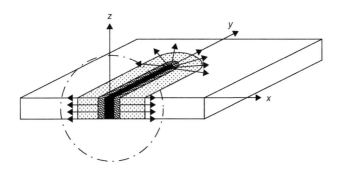

FIG. 4.12 Temperature distribution on butt-welded thin plates.

It should be noted that Eq. (4.7.1) is the theoretical solution of temperature distribution for the case where the heat source q_1 per unit length of a line (breadth $= 0$) is given on an infinitely large plate with unit thickness. In actual welding, the heat source distributes in a finite breadth along the weld line and the breadth of plate is also finite. Because of these facts, the solution given by Eq. (4.7.1) is regarded as an approximation.

As a simple exercise, let's show that the transient temperature distribution given by Eq. (4.7.1) satisfies the differential equation governing the one-dimensional heat conduction. Neglecting heat transfer from the surface, the heat conduction equation is expressed as

$$c\rho \frac{\partial T}{\partial t} = \lambda \frac{\partial^2 T}{\partial x^2}. \tag{4.7.2}$$

By substituting Eq. (4.7.1) into the left and right sides of Eq. (4.7.2), the following equations are obtained:

$$\frac{\partial T}{\partial t} = -\frac{1}{2t}T + \frac{x^2}{4kt^2}T \tag{4.7.3}$$

$$\frac{\lambda}{c\rho} \frac{\partial^2 T}{\partial x^2} = k\frac{\partial^2 T}{\partial x^2} = -\frac{1}{2t}T + \frac{x^2}{4kt^2}T. \tag{4.7.4}$$

From the above equations, it can be readily shown that Eq. (4.7.1) satisfies Eq. (4.7.2). Further, the equation for the maximum temperature in Table 4.4 can be derived from Eq. (4.7.1), noting that the temperature becomes the highest when $\partial T/\partial t = 0$, that is,

$$\frac{\partial T}{\partial t} = \frac{q_1}{c\rho}\left\{\frac{x^2}{4kt^2} - \frac{1}{2t}\right\}\frac{e^{-\frac{x^2}{4kt}}}{(4\pi kt)^{1/2}} = 0. \tag{4.7.5}$$

From Eq. (4.7.5), the time t when the temperature at point x reaches the maximum temperature is given as

$$t = \frac{x^2}{2k}. \tag{4.7.6}$$

By substituting t given by Eq. (4.7.6) into Eq. (4.7.1), the maximum temperature T_{max} is obtained as

$$T_{max} = T\left(x, \frac{x^2}{2k}\right) = \frac{q_1}{c\rho}\frac{1}{\sqrt{2\pi e}\,x}. \tag{4.7.7}$$

Further, Eq. (4.7.7) can be related to the welding problem. Assuming that the plate thickness is h and the welding current, the voltage, the speed, and the efficiency are I, U, v, and η, respectively, the heat input q_1 is given as

$$q_1 = Q_{net}/h = IV\eta/vh. \tag{4.7.8}$$

Assuming that $h = 5$ mm, $I = 200$ A, $U = 20$ V, $v = 5$ mm/s, and $\eta = 0.5$, for example, the maximum temperature at a distance r from the weld line $T_{max}(r)$

can be calculated using the following equation, where e is the base of natural logarithm ($e = 2.718$):

$$T_{max}(r) = \frac{q_1}{c\rho}\left(\frac{1}{2\pi e}\right)^{1/2}\frac{1}{r} =$$

$$\left(\frac{0.5 \times 200 \times 20}{5 \times 5 \times 0.47 \times 7.87 \times 10^{-3}}\right)\left(\frac{1}{2 \times 3.14 \times e}\right)^{1/2}\frac{1}{r} = \frac{5235}{r}. \tag{4.7.9}$$

To measure the temperature decreases with the distance from the weld line, calculate the maximum temperatures at 20 mm, 30 mm, and 40 mm away from the weld line as examples. It is found that $T_{max}(20) = 261.8°C$, $T_{max}(30) = 174.5°C$, and $T_{max}(40) = 130.9°C$ from Eq. (4.7.9). It should be noted that the initial temperature is assumed to be 0°C in this case. If the initial temperature is T_0, T_0 must be added to Eq. (4.7.7) to obtain the maximum temperature.

4.7.3 Temperature Distribution on a Butt-Welded Joint of a Thick Plate

A detailed explanation of the theory for two-dimensional problems has not been given so far. The theory is not so much different from one- or two-dimensional problem. In this section, their similarities and differences are shown using temperature distributions in thin and thick plates as examples.

Let's consider the temperature distribution due to a line weld (bead weld) laid on the surface of a thick plate placed on the x-y plane as shown in Fig. 4.13. When the speed of welding is very fast or when we are looking at the location far behind from the moving heat source, the heat source can be simplified as an instantaneous line heat source. In this case, the heat flows in the x and z directions simultaneously. If the heat transfer from the surface is neglected, the transient temperature distribution is given as

$$T(r, t) = \frac{q_2}{c\rho}\frac{e^{-\frac{r^2}{4kt}}}{4\pi kt}, \tag{4.7.10}$$

where $r = (x^2 + z^2)^{1/2}$.

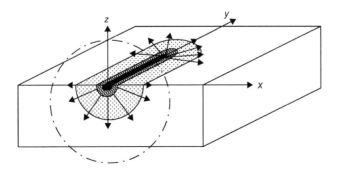

FIG. 4.13   Temperature distribution on butt-welded thick plates.

It can be readily shown that the temperature given by Eq. (4.7.10) satisfies the following equation for two-dimensional heat conduction:

$$c\rho \frac{\partial T}{\partial t} = \lambda \left(\frac{\partial^2 T}{\partial x^2} + \frac{\partial^2 T}{\partial z^2} \right). \qquad (4.7.11)$$

By substituting Eq. (4.7.10) into the left and the right sides of Eq. (4.7.11), the following equations are obtained:

$$\frac{\partial T}{\partial t} = -\frac{1}{t}T + \frac{r^2}{4kt^2}T \qquad (4.7.12)$$

$$\frac{\lambda}{c\rho} \left(\frac{\partial^2 T}{\partial x^2} + \frac{\partial^2 T}{\partial z^2} \right) = k\left(\frac{\partial^2 T}{\partial x^2} + \frac{\partial^2 T}{\partial z^2} \right) = -\frac{1}{t}T + \frac{x^2 + y^2}{4kt^2}T = -\frac{1}{t}T + \frac{r^2}{4kt^2}T. \qquad (4.7.13)$$

From these equations, it is seen that Eq. (4.7.10) satisfies Eq. (4.7.11).

In the same way as the thin plate, when a line heat source of Q_{net} per unit length is provided on the surface, the maximum temperature T_{max} is calculated by

$$T_{max} = T(r, \frac{r^2}{4k}) = \frac{2Q_{net}}{c\rho} \frac{1}{e\pi r^2}. \qquad (4.7.14)$$

If the same welding conditions as in the thin plate with 5mm thickness—that is, $I = 200$ A, $U = 20$ V, $v = 5$ mm/sec, and $\eta = 0.5$—are assumed, the maximum temperatures at 20 mm, 30 mm, and 40 mm away from the weld line are found to be $T_{max}(20) = 31.7°C$ and $T_{max}(30) = 14.1°C$, $T_{max}(40) = 7.9°C$, according to Eq. (4.7.14). Comparing this result with that of the thin plate, it is seen that the maximum temperature deceases rapidly with the distance from the weld line.

4.7.4 Inherent Strain Distribution on a Butt-Welded Joint of a Thin Plate

When a thin plate is welded, the maximum temperature T_{max} at a point distance r away from the weld line is inversely proportional to the distance, and it is given by

$$T_{max}(r) = 0.242 \frac{Q}{c\rho h} \frac{1}{r}. \qquad (4.7.15)$$

As discussed in Section 1.2, when a fully restrained bar is subjected to a thermal cycle $0 \to T_{max} \to 0$, the plastic strain ε^i and the residual stress σ are given as according to Table 1.4 in Chapter 1:

$$\begin{array}{lll}
\text{For } T_{max} \leq T_Y, & \varepsilon^i = 0 \\
& \sigma = 0 \\
\text{For } T_Y \leq T_{max} \leq 2T_Y, & \varepsilon^i = -\alpha(T_{max} - T_Y) \\
& \sigma = E\alpha(T_{max} - T_Y) & (4.7.16) \\
\text{For } 2T_Y \leq T_{max}, & \varepsilon^i = -\varepsilon_Y \\
& \sigma = \sigma_Y,
\end{array}$$

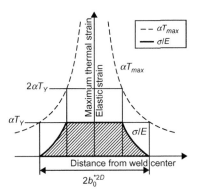

FIG. 4.14 Distribution of thermal strain and elastic strain in fully restrained plate after welding.

where

$$T_Y = \sigma_Y/E\alpha$$

$$\varepsilon_Y = \sigma_Y/E.$$

Noting that the restraint in the welding direction is close to the fully restrained state, the residual stress in the welding direction near the weld line can be approximated by Eq. (4.7.16). According to this equation, the residual stress distribution becomes trapezoidal-shaped, as shown in Fig. 4.14. It is also seen that the width of the plastic strain (inherent strain) region becomes $2b_0^{*2D}$, which is given by

$$2b_0^{*2D} = \frac{q_1}{c\rho} \frac{2}{\sqrt{2\pi e}\, T_Y} = \frac{q_1}{c\rho} \frac{2E\alpha}{\sqrt{2\pi e}\, \sigma_Y} = \frac{2E\alpha Q_{net}}{\sqrt{2\pi e}\, c\rho h\sigma_Y}. \tag{4.7.17}$$

Equation (4.7.17) is derived from the fact that the plastic deformation occurs when the temperature given by Eq. (4.7.9) is higher than $T_Y = \sigma_Y/E\alpha$. These characteristics may be confirmed by the analysis using the FEM, which will be discussed in Chapter 6. When the plate is thick, Eq. (4.7.14) can be used to obtain the width of the plastic strain region $2b_0^{*3D}$, and it is shown to be

$$2b_0^{*3D} = 2\left(\frac{2Q_{net}}{c\rho} \frac{1}{e\pi T_Y}\right) = 2\left(\frac{2Q_{net}}{c\rho} \frac{E\alpha}{e\pi\sigma}\right)^{1/2} = 2\left(\frac{2\alpha E Q_{net}}{e\pi c\rho\sigma_Y}\right)^{1/2}. \tag{4.7.18}$$

Further, by integrating the stress distribution given by Eq. (4.7.16) over the cross-section of the weld joint, the tendon force F_T is obtained. In a thin plate, the tendon force becomes

$$F_T = 0.335(E\alpha/c\rho)Q_{net}. \tag{4.7.19}$$

It should be noted that the stress in the welding direction must maintain equilibrium when the plate is free in the welding direction, compressive stress is produced in the area away from the welding where $T_{max} < T_Y$.

4.8 FLOW OF ANALYSIS FOR WELDING DEFORMATION AND RESIDUAL STRESS

Numerical simulation based on computational welding mechanics is a very effective means to understand problems related to welding deformation and residual stress and to obtaining suggestions for solving these problems. The steps for reaching solutions using numerical simulation are summarized in the form of a flowchart, as shown Fig. 4.15.

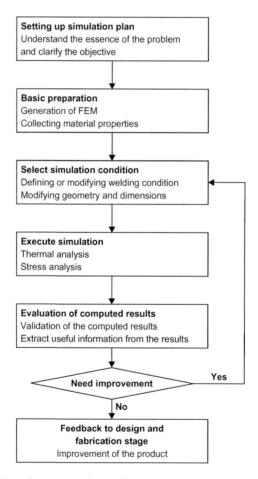

FIG. 4.15 Flowchart of process to solve problems using simulation.

Before starting the simulation, however, we have to first understand the problem well. Then the purpose of the simulation can be clearly defined.

Once the object to be analyzed and the purpose are fixed, we set up an analysis model and prepare input data. With these data, we conduct the heat conduction and stress analyses. From the computed deformation and stress, we need to check whether the input data and the procedure of the simulation are correct. If the computed results are reasonable compared to an available analytical solution or commonsense, we can proceed to analyze them to get quantitative information or hints to solve the problem. According to this, the simulation conditions, such as the welding conditions and the geometry of the structure, can be modified, and the simulation is repeated until improvement is confirmed. Table 4.6 describes all the steps of simulation, from the definition of the problem to obtaining the solution.

TABLE 4.6 Procedure to Carry Out Simulation in Steps

Set up simulation plan	Based on the judgment of whether the problem at hand is related to stress or deformation and which part of the structure is causing the problem, clarify the objective and purpose.
Basic preparation for simulation	Decide whether the problem can be simplified as a two-dimensional problem and model the structure as FEM. Collect information on temperature-dependent material properties in the form of tables or functions.
Select simulation condition	Define or modify the welding conditions, such as the welding current, voltage, speed, heat input distribution, heat input efficiency, initial temperature, interpass temperature, clamping and tack welding. If necessary, modify the geometry and the dimensions of the model.
Execute simulation	Compute temperature distribution with advancing the time by a given time increment. Using these temperature distributions, compute the deformation and the stress at each time step. When the structure cools down to room temperature the final deformation and the residual stress are obtained.
Evaluation of computed results	Validate the computed results with respect to the maximum values and distribution of temperature, stress, and strain by comparing them with melting temperature, yield stress, etc. Through this validation, careless errors in input data can be detected.
	Extract useful information from the results. If the deformation exceeds the allowable limit, for example, thickness can be increased or the welding heat input can be reduced.
Feedback to design and fabrication stage	The prediction obtained through simulation can be applied to design or fabrication procedure to improve products.

4.9 CHECKLIST FOR RATIONAL SIMULATION

4.9.1 Checklist for Preparation of Input Data

4.9.1.1 Confirmation of Necessary Information to be Supplied

For heat conduction analysis, thermal visco-elasto-plastic analysis, and inherent strain analysis, the information listed in Table 4.7 must be prepared according to the type of simulation to be done.

TABLE 4.7 Information Necessary When Preparing Input Data

Information necessary for FE mesh	Shape, dimensions, minimum and maximum element size, maximum number of elements and nodes allowed for the FEM code to be used; table of nodal coordinate and table of nodal point number of elements
Information necessary for thermal analysis	(Material properties) density, specific heat, heat conductivity, heat transfer coefficient (Initial/boundary condition) ambient temperature, initial temperature, location of surface facing atmosphere, prescribed temperature and its location (Welding condition) distribution and start/stop time of heat source in volume distribution and start/stop time of heat source on surface type of heat source (stationary/moving, moving speed) starting/finishing point, distribution (length, width) of heat source (Computing condition) time increment, maximum computing steps, or time to finish computation
Information necessary for thermal visco-elasto-plastic analysis	(Material properties) Young's modulus, Poison's ratio, yield stress, strain hardening coefficient, thermal expansion coefficient, creep properties, properties of groove before welding (Initial/boundary condition) fixed displacement, specified displacement, applied load (Computing condition) total steps of thermal data for input, allowable error, mechanical melting point, maximum temperature in stress analysis, maximum time increment, time increment for creep analysis
Information necessary for computation using inherent strain	(Material properties) Young's modulus, Poison's ratio (Initial/boundary condition) fixed displacement, specified displacement, applied load (Computing condition) choice between plane stress and plane strain, choice of element type (Forward analysis) inherent strain given to each element (Inverse analysis) location and component of inherent strain to be determined location and component of measured strain and their values

4.9.1.2 Consistent Use of Units

Before entering into calculation, make sure the units of input data are consistent. Otherwise, errors may appear in the decimal points of given material properties and computed stresses and deformation. To avoid such errors, you should use standard units, such as m for length and kg for the mass. In the field of welding engineering, the dimensions of welded joints are frequently specified in mm and cm, so the combination mm-kg or cm-g may be used. In such cases, it is best to use the consistent units listed in any column of Table 4.3. Special attention should be paid to decimal points of material properties and physical properties, which are shaded in the table. For example, when the combination of mm-kg is used, 1 appears in the output as displacement implies 1 mm = 0.001 m. Similarly, 1 in stress means $1 \text{ N/mm}^2 = 10^6 \text{ N/m}^2 = 10^6 \text{ Pa}$.

4.9.2 Checklists for the Results of Simulation

4.9.2.1 Checklist Applicable to Both Heat Conduction Analysis and Stress Analysis

- Confirmation of input data by comparing with data printed out.
- Checking the mesh division by viewing the FEM figure.
- Checking that no element is extremely distorted or has a large aspect ratio.
- Checking that the size of the element is small near the weld bead and smoothly becomes large with distance from the weld bead.

4.9.2.2 Checklist for Results of Heat Conduction Analysis

- Examine the temperature at the last step, which is given in the output file (heat2d.post) and check that the maximum temperature is higher than the melting point of the material (it should not be extremely high).
- The distribution of the maximum temperature should be smooth without irregularity.
- It should be confirmed that a sufficient number of elements (about three elements in three directions) are included in the weld pool where the highest temperature is higher than the melting temperature. If it is insufficient, the plastic strain concentrated near the weld bead will not be accurately computed.
- The temperature at the nodes after complete cooling should be the same as the ambient temperature. This can be checked by viewing the temperature at the last step given in the output file (temp.file).

4.9.2.3 Checklist for Results of Thermal Elastic-Plastic Analysis

- Confirm that the boundary conditions and the symmetry condition are correctly given by viewing the displacement u, v at the last step. Check that the maximum value of the stress is not unrealistically large, and there is no irregular distribution of the stress or the displacement.

- Check the equivalent plastic strain distribution. It must distribute only in the region near the weld bead, otherwise there may be an error in defining the boundary condition or the temperature data.
- Check that the equivalent stress at the last step is not too large compared to the yield stress at room temperature and that the stress normal to the free surface is zero or nearly zero.

4.9.2.4 Checklist for Results of Inherent Strain Analysis

- Confirm that the boundary conditions and the symmetry condition are correctly given by viewing the displacement u, v at the last step. Check that the maximum value of the displacement is not unrealistically large or small and that there is no irregular distribution of the displacement.
- When the residual stress is computed using the given inherent strain, confirm that there is no error in the distribution and the direction of the inherent strain given as the input data by viewing the stress distribution.
- In inverse analysis, confirm accuracy by comparing the elastic strain computed using the inherent strain estimated through the inverse analysis and the measured elastic strain given as the input data.

4.10 TROUBLESHOOTING FOR PROBLEMS EXPERIENCED IN COMPUTATION

4.10.1 Troubleshooting for Common Problems in Heat Conduction Analysis and Stress Analysis

There is disagreement between the input data and its printout.

- Check for duplication or missing data.
- Check that the order of the data is correct.
- Check that the number of nodal points and elements are given in sequential order.
- Check that the format of the input data is correct.
- Make sure that two-byte code is not included in the input file.

Jacobean at an integration point is zero or negative.

- Check that elements are not collapsed or the sequence of the node in the element given as the input is correct.
- Check the coordinates of the nodes.

4.10.2 Troubleshooting for Heat Conduction Analysis

The maximum temperature is below or greatly exceeding the melting temperature of the material.

- Check that the welding condition listed in Table 4.7 is correctly defined in the input data.

- Check that the material constants listed in Table 4.7 are correctly given. Check that the units of the data are correct using Table 4.8. Make sure that data, such as the density, thermal conductivity, and heat capacity, are correctly given according to the input manual.

TABLE 4.8 Units of Physical Values Used in Welding Mechanics

Physical Value	Symbol	Unit (Standard)	Using mm and kg	Using cm and g
Length	L	m	10^{-3} m = 1 mm	10^{-2} m = 1 cm
Force	F	N	N	N
Time	t	s	sec	sec
Velocity	v	m/sec	0.001 m/s = 1 mm/sec	0.01 m/s = 1 cm/sec
Temperature	T	°C, K	°C, K	°C, K
Mass	m	kg	kg	10^{-3} kg = 1g
Voltage	U	V	V	V
Current	I	A	A	A
Strain	ε	nondimension	nondimension	nondimension
Stress	σ	Pa = N/m²	10^6 Pa = 1 N/mm²	10^4 Pa = 1 N/cm²
Young's modulus	E	Pa = N/m²	10^6 Pa = 1 N/mm²	10^4 Pa = 1 N/cm²
Poison's ratio	ν	nondimension	nondimension	nondimension
Energy	W	J = Nm = AVs 1 cal = 4.186 J	10^{-3} J = 1 Nmm	10^{-2} J = 1 Ncm
Welding heat input	Q	J/m	10^3 J/m = 1 J/mm	10^2 J/m = 1 J/cm
Density	ρ	kg/m³	10^9 kg/m³ = 1 kg/mm³	10^3 kg/m³ = 1 g/cm³
Specific heat	c	J/kg°C	J/kg°C	10^3 J/kg°C = 1 J/g°C
Thermal expansion coefficient	α	1/°C	1/°C	1/°C
Thermal conductivity	λ	J/(sec m°C)	10^3 J/(sec m°C) = 1 J/(sec mm°C)	10^2 J/(sec m°C) = 1 J/(s cm°C)
Heat transfer coefficient	β	J/(sec m²°C)	10^6 J/(sec m²°C) = 1 J/(sec mm²°C)	10^4 J/(sec m²°C) = 1 J/(sec cm²°C)

The distribution of maximum temperature is not smooth.

- Check that the time increment is small enough.

The region above the melting temperature is too small compared to the expected size of the weld bead.

- Make sure that the size of the elements near the weld line is small enough (a large element means large heat capacity and low temperature).

The temperature after complete cooling (temp.file) is not the atmospheric temperature.

- The computation may be terminated because the computing step exceeds the maximum step specified in the input data. In this case, a sufficiently large maximum step must be given as the input data.
- The atmospheric temperature in the input data is not correctly defined.

The location of the maximum temperature just before complete cooling in the output file (heat2d.post) is different from the expected location.

- Check the surface subjected to heat transfer defined in the input data.

4.10.3 Troubleshooting for Thermal Elastic-Plastic Analysis

The displacements are not consistent with the given boundary condition and symmetry condition or their distribution is not smooth.

- Check that the boundary conditions and forces listed in Table 4.7 are correctly given.

Plastic strain is observed in the place not expected or the distribution of stress is not smooth.

- Check that the boundary conditions and forces listed in Table 4.7 are correctly given.

The equivalent stress at complete cooling is much larger than the yield stress of the material.

- The time increment given as the input data may be too large.

The computation does not converge at the first step or the first stage.

- Check that the material constants given as input data are correct and that the rigid body motion is fixed by the boundary condition.

The computation does not converge in the middle stage of the welding.

- Compute again with a small time increment or small temperature increment.

REFERENCES

[1] Courant R. Variational methods for the solution of problems of equilibrium and vibrations. Bull Am Math Soc 1943;49:1.

[2] Prager W, Synge J. Approximations in elasticity based on the concept of function spaces. Quart Appl Math 1947;5:241.

[3] Argyris J. Energy theorem and structural analysis. Aircraft Eng 1954;26:347.

[4] Turner M, Clough R, Martin H, Topp L. Stiffness and deflection analysis of complex structures. J Aeron Sci 1956;23:805.

[5] Clough R. FEM in plane stress analysis: Proc. 2nd Conf. Electronic Computation. Pittsburg: ASCE; 1960.

[6] Ueda Y, Yamakawa T. Analysis of thermal elastic-plastic stress and strain during welding by finite element method. Trans JWS 1971;2(2):90–100.

[7] Satoh K, Ueda Y, Fujimoto T. Welding deformation and residual stress. Tokyo: Sanpou Shupan (in Japanese); 1979.

[8] Yagawa G. Introduction to finite element method to flow and heat conduction analyses, series on basic and application of finite element method (8). Tokyo: Baifukan (in Japanese); 1983.

[9] Yagawa G, Miyamoto N. Finite element method for thermal stress, creep and heat conduction analyses. Tokyo: Saiensu-sha (in Japanese); 1985.

[10] Ueda Y, Nakacho K, Kim Y, Murakawa H. Fundamentals of thermal-elastic-plastic-creep analysis and measurement of welding residual stress for numerical analysis in thermal conduction analysis. Jl JWS 1986a;55(6):336–48 (in Japanese).

[11] Ueda Y, Nakacho K, Kim Y, Murakawa H. Fundamentals of thermal-elastic-plastic-creep analysis and measurement of welding residual stress for numerical analysis in thermal-elastic-plastic-creep analysis. Jl JWS 1986b;55(7):399–410 (in Japanese).

Chapter 5

Q&A for FEM Programs

The following programs are provided on this book's companion website (herein, website):

1. Welding heat conduction FEM program.
2. Welding thermal elastic-plastic creep FEM program.
3. Inherent strain FEM program for measurement and prediction of welding residual stress.
4. Simple graphical postprocessing program.

These programs are supplied only to the readers of this book. It is against the law to copy and/or sell them.

The programs work only on Windows XP or Windows 7 machines. The recommended memory is 2 GB. The total number of nodes and elements are limited to 10,000. The authors cannot provide program support, and users are responsible for the results obtained by the programs.

5.1 Q&A FOR PROGRAM INTRODUCTION

Q1: What programs are offered on the website?

A1: Four executable programs as shown in Table 5.1 are given in the folder PROG:

1. **heat2d.exe** is a two-dimensional welding heat conduction FEM program that can compute temperature distribution during welding and cooling.

   ※**heat2d** is the abbreviated name of the two-dimensional program for heat conduction.

2. **tepc2d.exe** is a two-dimensional welding thermal elastic-plastic creep FEM program that can compute transient and residual welding deformation and stresses.

   ※**tecp2d** is the abbreviated name of the two-dimensional program for thermal elastic-plastic creep.

3. **inhs2d.exe** is a two-dimensional computational measuring program for welding residual stresses and deformation based on the inherent strain method.

   ※**inhs2d** is the abbreviated name of the two-dimensional program for inherent strain.

4. **awsd.exe** is a simple postprocessor to show results (i.e., temperature, displacement, strain, and stress) graphically.

   ※**awsd** is the abbreviated name of the postprocessor for analysis of welding structure deformation.

   #heat2d.exe, tepc2d.exe and inhs2d.exe were developed by the authors, and awsd.exe was developed by Professor Yu Luo.

Q2: What computed results can be obtained using these programs?

A2: The computed results shown in Table 5.2 can be obtained and graphically displayed.

TABLE 5.1 Program List in PROG

Welding heat conduction 2D FEM program heat2d.exe	Thermal elastic-plastic creep 2D FEM program tepc2d.exe
Computational measuring 2D inherent strain program for residual stress and deformation inhs2d.exe	Simple postprocessing program awsd.exe

---

TABLE 5.2 Program Features in Computation and Graphical Representation

Welding heat conduction program	• Temperature distribution • Max. temperature distribution • Heating rate and cooling rate • Melting zone • Heat affected zone • Mechanical melting zone
Welding thermal elastic-plastic creep program	• Plastic strain zone • Elastic strain zone • Transient stress and strain • Residual stress and strain • Creep strain • Inherent strain
Inherent strain program for stress measurement	• Residual stresses • Deformation • Inherent strain • Measurement • Prediction
Simple postprocessing program	• Temperature display • Temperature rate display • Deformation display • Stress display • Inherent strain display • Measured strain display

As examples, the postprocessing program **awsd.exe** can display temperature distribution (Fig. 5.1), welding deformation (Fig. 5.2), and residual stress (Fig. 5.3).

Q3: How do you install the programs?

A3: Programs can be easily installed by copying and pasting as follows:

1. Copy programs **heat2d.exe**, **tepc2d.exe**, and **inhs2d.exe** in **PROG** to your data folders for computation.
2. Copy postprocessing program **awsd.exe** in **PROG** to the desktop of your computer for convenient operation.
3. Create a folder **JWRI\LICENSE** and then copy license file **JWRI\ License\JWRI_FILE** to the folder **JWRI\LICENSE**.

Q4: How do you use the programs?

A4: Readers can use the welding heat conduction program **heat2d.exe**, the thermal elastic-plastic creep program **tepc2d.exe**, and the inherent strain program **inhs2d.exe** using the following steps:

1. Decide what program you will use according to your purpose: **heat2d. exe**, or **tepc2d.exe**, or **inhs2d.exe**.

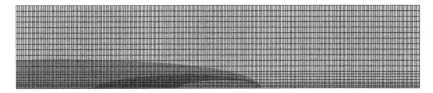

FIG. 5.1 Transient temperature distribution during butt welding (half model).

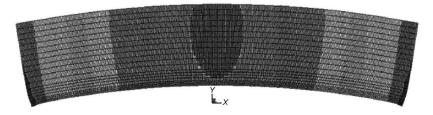

FIG. 5.2 Residual in-plane bending deformation due to thermal cutting.

FIG. 5.3 Transverse residual stress σ_y distribution in butt weld (half model).

2. Copy the **.exe** file from folder **PROG** to your data folder, for example, **DATA**.

3. Prepare input file with extension *****.inp** in your data folder **DATA** following the direction given in the user manuals on the book's website. The format of input data is not supported by pre-processors yet and readers have to prepare new input data manually.

4. For practice purposes, readers do not need to create new input data themselves. We recommend you download the example files *****.inp** saved in the folders **DATA1-DATA21** on the book's website to your own folder. The input files named by *****.inp** is **heat2d.inp** or **tepc2d.inp** or **inhs2d.inp**.

5. Double-click the program in your data folder and start computing.

The filenames of the input data and the details for input data, program, name and output filenames, postprocessing program name, and input filenames for the postprocessor are given in Fig. 5.4 for the welding heat conduction program **heat2d.exe**. Similar steps are described by Figs 5.5 and 5.6 for the thermal elastic-plastic creep program **tepc2d.exe** and the inherent strain–based residual stress measuring program **inhs2d.exe**.

FIG. 5.4   Steps for using welding heat conduction program **heat2d.exe**.

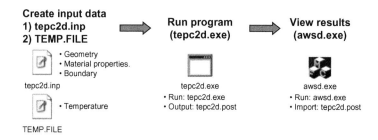

FIG. 5.5   Steps for using welding thermal elastic-plastic creep program **tepc2d.exe**.

FIG. 5.6   Steps for using inherent strain program for residual stress measurement **inhs2d.exe**.

5.2 Q&A FOR WELDING HEAT CONDUCTION PROGRAM heat2d.exe

Q1: What functions and limitations does the program **heat2d.exe** have?

A1: Program **heat2d.exe** has the following functions and limitations:

1. Transient welding heat conduction behaviors can be computed.
2. Four-node isoparametric plane element with full Gaussian integration points are used in the program.
3. Temperature-dependent material properties (i.e., density, thermal conductivity, specific heat, heat transfer coefficient) are defined by multipoints (Ti, Pi), where Ti and Pi are temperature and material properties at Ti, respectively. The material properties at an arbitrary temperature are calculated in the program by interpolation or extrapolation.
4. Initial temperature and fixed temperature boundaries can be defined. Heat radiation surfaces and heat convection surfaces are assumed to be the same in the program where heat flow between the boundary and atmosphere is modeled by the heat transfer coefficient.
5. The welding heat source is modeled by both internal heat generation and surface heat flux.
6. The welding heat source can move in a straight weld line or be fixed in a certain area.
7. Besides temperature distribution, the maximum temperature in thermal cycles, the rate of temperature change, and the average rate of temperature change between two specified temperatures in the thermal cycle can be computed.

Q2: What is the model for a moving welding heat source?

A2: A moving welding heat source is a source of heat that moves along the welding line, and metal on the welding line is heated in sequence.

   The distribution of the welding heat source is very important to predict the melting pool during welding. It has relatively little effect on welding deformation and residual stress. The most popular distribution model is described by the Gaussian function, and the double elliptical distribution model proposed by Goldak [1] has a higher accuracy.

   The welding heat source model used in the program **heat2d.exe** is idealized and has uniform distribution in a rectangle prism, as shown in Fig. 5.7. The length, width, and depth of the rectangle heat source are denoted by a, b, and h, respectively.

   If welding current I, voltage U, and welding speed V of arc welding are known, and heat efficiency η, size parameters (length a, width b, depth h, equal to plate thickness) of a moving heat source are determined by experience or experiment. The heat generation rate per volume $\dot{q}_V$ [J/(mm^{3}sec)]

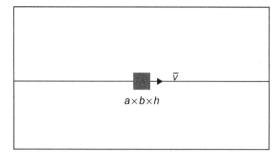

FIG. 5.7   Schematic representation of moving heat source model.

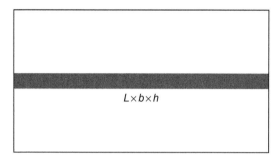

FIG. 5.8   Schematic representation of fixed heat source model.

and surface heat flux $\dot{q}_A$ [J/(mm²sec)] can be calculated by Eqs. (5.2.1) and (5.2.2):

$$\dot{q}_V = \eta \frac{I \times U}{a \times b \times h} \tag{5.2.1}$$

$$\dot{q}_A = \eta \frac{I \times U}{a \times b} . \tag{5.2.2}$$

Q3: What is the model for a fixed welding heat source?

A3: A fixed welding heat source is such a computational model as all metal in a weld zone is heated together within a certain period of time. The size parameters of the weld zone for a butt joint are weld length L, weld width b, and weld depth h, as shown in Fig. 5.8.

If the heating time of the fixed heating source model is assumed to be t_w and the welding current I, voltage U, and welding speed V of arc welding are known, heat generation rate per volume $\dot{q}_V$ [J/(mm³sec)] and surface heat flux $\dot{q}_A$ [J/(mm²sec)] can be calculated by Eqs. (5.2.3) and (5.2.4), respectively:

$$\dot{q}_V = \eta \frac{I \times U}{L \times b \times h} = \eta \frac{I \times U}{V \times b \times h} \cdot \frac{1}{t_w} \tag{5.2.3}$$

$$\dot{q}_A = \eta \frac{I \times U}{L \times b} = \eta \frac{I \times U}{V \times b} \cdot \frac{1}{t_w}, \qquad (5.2.4)$$

where $L (= V \times t_w)$, b, and h (h is the same as plate thickness) in Eq. (5.2.3) to Eq. (5.2.4) are the length, width, and depth, respectively, of the fixed heat source.

The heating time t_w of a fixed heat source model may be the actual welding time theoretically and is often taken as an input variable in numerical simulation. If heating time t_w is long, the heat generation rate per volume $\dot{q}_V$ or surface heat flux $\dot{q}_A$ will be small and then the temperature in the heating process may increase very slowly. If the heating time t_w is too short, the temperature in the heating zone may become too high. In actual numerical simulation, the heating time t_w is adjusted to fit the maximum temperatures in the welding thermal cycle and the width of melting temperature zone or mechanical melting temperature zone. The heating time t_w with 1 sec is often used.

By the way, the fixed welding heat source model is mainly used to simulate the macrothermal mechanical behaviors more than the local welding pool.

Q4: How do you define heat transfer surfaces?

A4: In the program, heat transfer surfaces are defined as follows:

As shown in Fig. 5.9, a plane element with four nodes ($p1$, $p2$, $p3$, $p4$) has normal surface $F5$ and back surface $F6$. The edge surfaces of the element are denoted by $F1$, $F2$, $F3$, and $F4$. The element heat transfer surfaces and heat insulation surfaces can be defined by a flag with 1 and 0, respectively. If an edge surface of the element is not connected with other elements, the edge surface will be automatically judged as a boundary surface in the program. If the boundary surface flag is 1, the heat flow from the boundary surface to the atmosphere will occur in the computation.

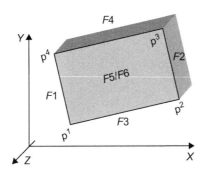

FIG. 5.9 Element surfaces.

Q5: How do you decide the time step for heat conduction computation?

A5: The initial time step or initial time increment must be set by the user in the input data. The maximum time step must satisfy Eq. (3.1.36) and Eq. (3.1.40) described in Section 3.1. When the program is executed, time step Δt will be automatically determined by the maximum temperature increment ΔT_{max} at the previous step. If ΔT_{max} at the previous step is larger than 20, the new time step Δt will be scaled smaller to make ΔT_{max} equal to 20. If ΔT_{max} at the previous step is less than 20, the new time step Δt will be scaled larger. When the maximum temperature over all the nodes is equal to the atmosphere temperature or the total step is equal to the input maximum steps, normally computation will be terminated.

Q6: What are the definitions of the rate of temperature change and the average rate of temperature change?

A6: The rate of temperature change is defined by Eq. (5.2.5) and is based on the temperature change at a small time increment. In the heating process, the rate of temperature change will be larger than 0. In the cooling process, the rate of temperature change will be less than 0. The average rate of temperature change is calculated based on the two specified temperatures $TC1$ and $TC2$, as shown in Fig. 5.10. According to the value of $TC1$ and $TC2$, the average rate of temperature change in the heating process or the cooling process can be automatically judged by Eq. (5.2.6) and Eq. (5.2.7), respectively. The time of $th1$ and $th2$ in the heating process or $tc1$ and $tc2$ in the cooling process is calculated from the two temperatures $TC1$ and $TC2$ defined in the input data.

Rate of temperature change:

$$\dot{T}(t) = dT/dt. \tag{5.2.5}$$

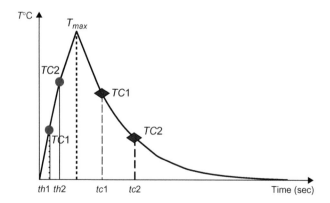

FIG. 5.10   Thermal cycle and average rate of temperature change between $TC1$ and $TC2$.

Average rate of temperature change (if $TC1 < TC2$):

$$\dot{T}_{av}(TC1,TC2) = \frac{TC2 - TC1}{th2 - th1} > 0. \qquad (5.2.6)$$

Average rate of temperature change (if $TC1 > TC2$):

$$\dot{T}_{av}(TC1,TC2) = \frac{TC2 - TC1}{tc2 - tc1} < 0. \qquad (5.2.7)$$

Q7: What files should be prepared before computation?

A7: The two files **heat2d.exe** and **heat2d.inp** in your computation folder and shown in Fig. 5.11 must be prepared before computation. The program module **heat2d.exe** must be copied to the computation folder from the book's website. The input file **heat2d.inp** must be written in the format following the user manual, **Manual\heat2d_manual.pdf**. The input file **heat2d.inp** is not supported by pre-processors yet and readers have to prepare the new input data manually based on the user manual on the book's website.

Q8: What output files are saved after computation?

A8: If the computation terminates normally, the results will be saved in the following files, as shown in Fig. 5.12:

1. **check.txt:** The output file to check input data.
2. If the input data has errors, they will occur after the program **heat2d. exe** starts running. You can check the file near the last several lines and guess where the error occurs in the input data.
3. **tnmax.txt:** The output file saving the maximum temperature of the thermal cycle at all nodes.
4. **TEMP.FILE:** The output file saving the temperature data for all nodes at all steps.

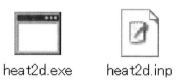

heat2d.exe   heat2d.inp

FIG. 5.11   Files in data folder before heat conduction computation.

check.txt   tnmax.txt   TEMP.FILE   heat2d.post

FIG. 5.12   Output files after heat conduction computation.

※**TEMP.FILE** will be the input file for the thermal elastic-plastic creep program **tepc2d.exe**.

5. **heat2d.post:** The output file for postprocessor **awsd.exe**.

6. A detailed format of the output file **heat2d.post** is given in the user manual, **Manual\heat2d_manual.pdf**.

5.3 Q&A FOR THERMAL ELASTIC-PLASTIC CREEP PROGRAM tepc2d.exe

Q1: What functions and limitations does program **tepc2d.exe** have?

A1: Program **tepc2d.exe** has the following functions and limitations:

1. This program can only analyze the thermal elastic-plastic creep behaviors at the plane stress state.

2. An isoparametric plane element with four nodes and full Gaussian integration points is used in this program.

3. Four types of loads, which are equivalent to nodal force, node displacement, thermal strain, and creep strain, can be used in this program.

4. The behaviors of groove or gap to be welded is modeled by a nonstiffness dummy element before welding. When the groove is filled with melted weld metal and when its temperature is cooled to its mechanical melting temperature, the dummy element becomes a normal element.

5. Temperature-dependent material properties (e.g., thermal expansion coefficient, Young's modulus, Poisson's ratio, yield stress, and work hardening tangent) are defined by multipoints (Ti, Pi), where Ti and Pi are temperature and material properties at Ti, respectively. The material properties at an arbitrary temperature are calculated in the program by interpolation or extrapolation.

6. The temperature increment or time increment can be set by input data. If the difference between the internal equivalent nodal force and the external equivalent nodal force is larger than convergence tolerance, the temperature increment or time increment will be automatically reduced to half until the convergent solution is obtained.

Q2: What files should be prepared before computation?

A2: The three files **tepc2d.exe**, **tepc2d.inp**, and **TEMP.FILE** in your computation folder and shown in Fig. 5.13 must exist before computation.

tepc2d.exe   tepc2d.inp   TEMP.FILE

FIG. 5.13 Files in data folder before thermal elastic-plastic creep computation.

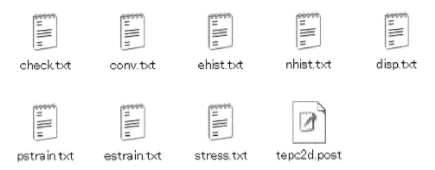

FIG. 5.14 Output files after thermal elastic-plastic creep computation.

The program module **tepc2d.exe** must be copied to the computation folder from the book's website. Element information, material properties, boundary conditions, and the computation control parameters must be defined in the input file **tepc2d.inp**, following the user manual, **Manual\tepc2d_manual.pdf**. The nodal temperatures at all time steps must be saved in **TEMP.FILE**, which has the same format as the output file **TEMP.FILE** after the heat conduction program **heat2d.exe** is executed.

Q3: What output files are saved after computation?

A3: If computation terminates normally, the results will be saved in the following files, as shown in Fig. 5.14:

1. **check.txt:** The output file to check input data. If the input data has errors, they will happen after the program **tepc2d.exe** starts running. Users can check the file near the last lines and guess where the error occurred in the input data.
2. **conv.txt:** The output file of the convergence information in computation.
3. **ehist.txt:** The output file for the change of variables (strain, stress, and temperature) at the given elements.
4. **nhist.txt:** The output file for the change of variables (displacement and temperature) at the given nodes.
5. **disp.txt:** The output file of the residual displacements at all nodes.
6. **pstrain.txt:** The output file of the residual plastic strains at all elements.
7. **estrain.txt:** The output file of the residual elastic strains at all elements.
8. **stress.txt:** The output file of residual stresses at all elements.
9. **tepc2d.post:** The output file for postprocessor **awsd.exe**. A detailed format of the output file **tepc2d.post** is given in the user manual, **Manual\tepc2d_manual.pdf**.

5.4 Q&A FOR THE INHERENT STRAIN–BASED PROGRAM inhs2d.exe

Q1: What functions and limitations does program **inhs2d.exe** have?
A1: Program **inhs2d.exe** has the following functions and limitations:

1. Predicts residual stresses when inherent strains are known.
2. Inversely computes inherent strains from the measured strains.
3. Inversely computes inherent strains from the measured strains and then computes residual stress distributions.
4. Residual stresses at the plane stress state and generalized plane strain state can be computed.
5. A four-node isoparametric plane element with full Gaussian integration points is used in this program.

Q2: What is the plane stress state?
A2: In the plane stress state only in-plane stress components exist and out-of-plane stress components do not exist. For example, if the x-y plane is the in-plane of a plate, the stresses $\sigma_x, \sigma_y, \tau_{xy}$ exist and the out-of-plane stresses $\sigma_z, \tau_{xz}, \tau_{yz}$ are zero.

Q3: What are the plane strain state and the generalized plane strain state?
A3: When the strain in the normal direction of a plane is zero, it is called the plane strain state. For example, if two ends of a weld line are constrained during welding, the strain in the welding direction should be zero, and it is in a plane strain state if the transverse section is considered as an object plane.

When the strain ε_z in the normal direction of an x-y plane (e.g., the transverse section plane of a long weld line) is assumed to distribute in the plane linearly defined by Eq. (5.4.1), the state is called a generalized plane strain state in this book:

$$\varepsilon_z(x,y) = A_1 + A_2 x + A_3 y. \tag{5.4.1}$$

In a generalized plane strain state, the section force Fz and the section moments My, Mx produced by normal stress σ_z must satisfy the following mechanical equilibrium equations:

$$
\begin{aligned}
Fz &= \int \left(E\varepsilon_z(x,y) - E\varepsilon_z^*(x,y) \right) dxdy \\
&= \int \left(E(A_1 + A_2 x + A_3 y) - E\varepsilon_z^*(x,y) \right) dxdy = 0
\end{aligned} \tag{5.4.2}
$$

$$
\begin{aligned}
My &= \int \left(E\varepsilon_z(x,y) - E\varepsilon_z^*(x,y) \right) xdxdy \\
&= \int \left(E(A_1 + A_2 x + A_3 y) - E\varepsilon_z^*(x,y) \right) xdxdy = 0
\end{aligned} \tag{5.4.3}
$$

$$Mx = \int \left(E\varepsilon_z(x,y) - E\varepsilon_z^*(x,y) \right) y \, dx \, dy$$
$$= \int \left(E(A_1 + A_2 x + A_3 y) - E\varepsilon_z^*(x,y) \right) y \, dx \, dy = 0. \tag{5.4.4}$$

Q4: What files should be prepared before computation?

A4: The two files **inhs2d.exe** and **inhs2d.inp** must exist in your data folder, as shown in Fig. 5.15, before computation. The program module **inhs2d.exe** must be copied to the computation folder from the book's website. The input file **inhs2d.inp** must be written in the format following the user manual, **Manual\inhs2d_manual.pdf**. The new input data have to be prepared by Microsoft Office or Microsoft WordPad because it is not supported by pre-processors yet.

Q5: What output files are saved after computation?

A5: If computation terminates normally, the results will be saved in the following files, as shown in Fig. 5.16:

1. **check.txt:** The output file to check input data. If the input data includes incorrect data, the error stop will happen after the program **inhs2d.exe** starts running. Users can check the file near the last several lines and guess where the wrong data is included in the input data.
2. **disp.txt:** The output file of residual displacements at all nodes.
3. **istrain.txt:** The output file of inherent strains at all elements.
4. **estrain.txt:** The output file of elastic strains at all elements.
5. **stress.txt:** The output file of residual stresses at all elements.
6. **inhs2d.post:** The output file for postprocessor **awsd.exe**. A detailed format of the output file **inhs2d.post** is given in the user manual, **Manual\inhs2d_manual.pdf**.

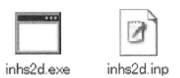

inhs2d.exe   inhs2d.inp

FIG. 5.15 Files in data folder before executing inherent strain program.

check.txt   disp.txt   istrain.txt   estrain.txt   stress.txt   inhs2d.post

FIG. 5.16 Output files after executing inherent strain program.

5.5 Q&A FOR POSTPROCESSING PROGRAM awsd.exe

Q1: What functions does postprocessing program **awsd.exe** have?

A1: Variables such as temperature, displacement, strain and strain at nodes, and elements can be graphically displayed by this program.

Q2: How do you use the postprocessing program **awsd.exe**?

A2: The postprocessing program can be used as follows:

1. To start the postprocessing program, first double-click the **awsd.exe** program icon on the desktop.
2. Push **left button** of mouse and then the postprocessing menu dialog shown in Fig. 5.17 will be displayed for importing result file.
3. To import the result file, go to **Post→View result→Select result file** (**heat2d.post** or **tepc2d.post** or **inhs2d.post**) from your **data folder→ Open**.

As an example, **DATA9\tepc2d.post** is shown in Fig. 5.18. The FEM model shown in Fig. 5.19 will be displayed after opening.

Mouse operation and function buttons:

- Translate: move the mouse while pushing the left mouse button.
- Rotation: move the mouse while pushing the right mouse button.

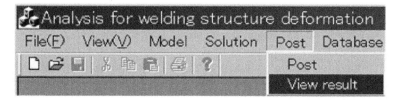

FIG. 5.17   Postprocessing menu dialog of **awsd.exe.**

FIG. 5.18   Dialog to select and open the result file.

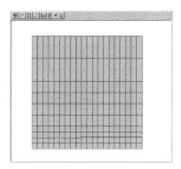

FIG. 5.19 FEM model displayed after post file is imported.

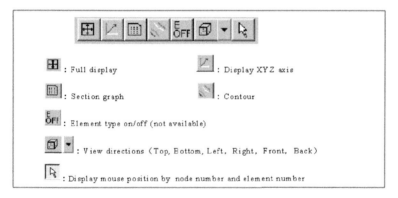

FIG. 5.20 Function dialog and buttons.

- Zooming: move the mouse while pushing the Ctrl key and the left mouse button.
- Figure 5.20 is the function buttons of the postprocessor.

To draw contours:

Figure 5.21 is the contour drawing dialog. The steps for drawing contours are as follows:

1. Select the variable from the list of DEF, DIS, REA, STN, STR, TEM.
2. Select the variable component from **direction** X or Y or Z. The corresponding relation between variables **DEF, DIS, REA, STN, STR, TEM** listed in the contour dialog and results respectively computed by program **heat2d.exe** or **tepc2d.exe** or **inhs2d.exe** are shown in Table 5.3.

To draw a graph:

The steps for drawing a graph are as follows:

1. Select the **Graph** button in function dialog and then click the left button of mouse

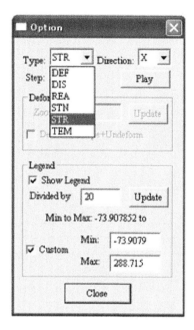

FIG. 5.21 Contour drawing dialog.

TABLE 5.3 Relation Between Variable in Contour Dialog and Result Computed by Programs

Awsd.exe	heat2d.exe	tepc2d.exe	inhs2d.exe
Contour variables	heat2d.post	tepc2d.post	inhs2d.post
DEF (Deformed shape)	—	Node displacement	Node displacement
DIS (Initial shape)	—	Node displacement	Node displacement
REA	Rate of temperature change or Average cooling rate (last step)	—	—
STN (%)	—	Plastic strain	Step 1: Measured strain Step 2: Inherent strain
STR	—	Residual stress	Step 1: Measured stress Step 2: Residual stress
TEM	Temperature or maximum temperature (last step)	Temperature or maximum temperature (last step)	—

2. Select a variable to be drawn from **Node Stress** or **Node Strain** or **Node Disp** and so on.
3. Select the variable component from **Draw (X)** or **Draw (Y)**.
4. Select all required nodes in sequence by **moving mouse point to a node** and **clicking**.

   The **Select** button at the bottom left of the dialog can be used to select nodes automatically by inputting the minimum node number, maximum node number, and number increment.

   The **Clear** button at the right of the Select button can be used to clear all selected nodes.
5. Click the **Save** button to save the selected nodes and variable values to a file. The saved file can be imported into Microsoft Excel.
6. Click **Draw Graph** and then a graph can be drawn, as shown in Fig. 5.22.

   The circle marks in the graph are the original value of the variable and the line is the curve approximated by the polynomial function.
7. Change the graph's horizontal axis from node number to a coordinate of *X* or *Y* or *Z* and a coordinate versus variable graph can be drawn.

   Figure 5.22 shows the graph drawing dialog and operation steps.

Q3: How do you draw a temperature contour?

A3: A temperature contour can be drawn using the following steps, as shown in Fig. 5.23:

1. Import the result file (e.g., **DATA8\heat2d.post**) after heat conduction computation.
2. Select TEM from Type list.
3. Select step number from Step.
4. If the last step of TEM is selected, the displayed temperature will be the maximum temperature in the thermal cycle.
5. Switch on **Show Legend and Custom** to adjust the range of contour.

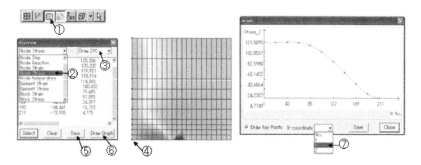

FIG. 5.22 Graph drawing dialog and drawing steps.

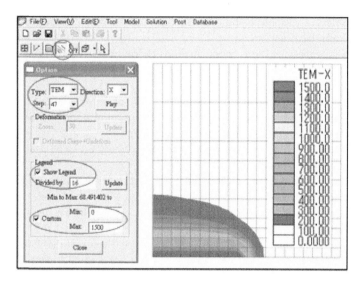

FIG. 5.23   Steps to draw temperature contour and maximum temperature distribution.

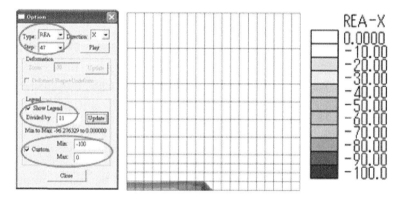

FIG. 5.24   Steps to draw the rate of temperature change and the average cooling rate between two specified temperatures.

Q4: How do you a draw cooling rate contour?

A4: A cooling rate contour can be drawn using the following steps, as shown in Fig. 5.24:

1. Import the result file (e.g., **DATA8\heat2d.post**) after the heat conduction computation.
2. Select REA from Type list.
3. Select step number from Step.
4. Switch on **Show Legend and Custom** to adjust the range of contour. REA > 0: Rate of temperature change in heating process (heating rate).

REA < 0: Rate of temperature change in cooling process (cooling rate). If the last step of REA is selected, the displayed rate of temperature change will be the average rate of temperature change or the average cooling rate between the two specified temperatures (e.g., 800–500°C) in the input data **heat2d.inp**.

Q5: How do you a draw stress contour?

A5: A stress contour can be drawn using the following steps, as shown in Fig. 5.25:

1. Import the result file (e.g., **DATA9\tepc2d.post**) after executing program **tepc2d.exe** or **DATA10\inhs2d.post** after executing program **inhs2d.exe**.
2. Select STR from Type list.
3. Select step number from Step.
4. Select the stress component from **direction** *X* or *Y*.
5. Switch on **Show Legend and Custom** to adjust the range of the contour. If the last step of STR is selected, the displayed stress will be the residual stress.

Q6: How do you draw a plastic strain contour?

A6: A plastic strain contour can be drawn using the following steps, as shown in Fig. 5.26:

1. Import the result file **tepc2d.post** (e.g., **DATA9\tepc2d.post**) after executing program **tepc2d.exe**.
2. Select STN from Type list.

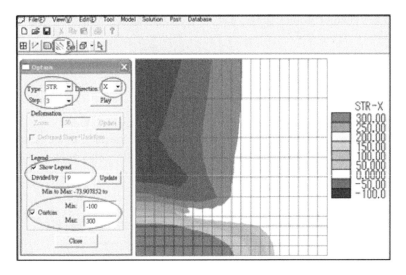

FIG. 5.25 Steps to draw stress contour and stress σ_x distribution.

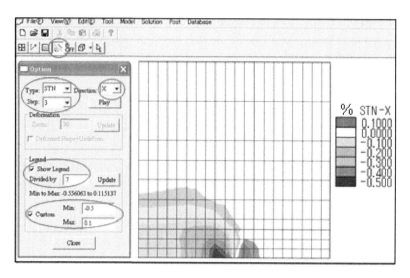

FIG. 5.26   Steps to draw plastic strain contour and contour of $\varepsilon_x^p(\%)$.

3. Select step number from Step.
4. Select the plastic strain component from **direction** X or Y.
5. Switch on **Show Legend and Custom** to adjust the range of the contour. If the last step of STN is selected, the displayed strain will be the residual plastic strain.
   The plastic strain displayed in **awsd.exe** is drawn in percentages (%).

Q7: How do you draw the inherent strain contour?

A7: The inherent strain contour can be drawn using the following steps, as shown in Fig. 5.27:

1. Import the result file **inhs2d.post** (e.g., **DATA10\inhs2d.post**) after executing program **inhs2d.exe**.
2. Select STN from Type list.
3. Select step number from Step.
4. Select the inherent strain component from **direction** X or Y.
5. Switch on **Show Legend and Custom** to adjust the range of the contour. **STN (%) at step 1** is the measured strain of the existing input data. (Zero value in nonmeasured elements.)
   STN (%) at step 2 is inherent strain.

Q8: How do you draw a stress distribution graph?

A8: A stress distribution graph can be drawn by the following steps, as shown in Fig. 5.28:

1. Click the **Graph** button.
2. Select a variable to be drawn from **Node Stress**.

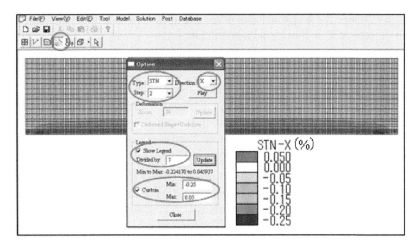

FIG. 5.27　Steps to draw inherent strain contour and contour of $\varepsilon_x^*(\%)$.

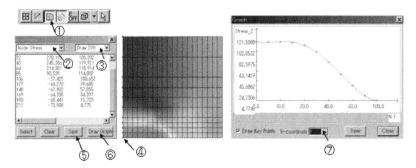

FIG. 5.28　Steps to draw the graph of stress distribution.

3. Select variable component from **Draw (X)** or **Draw (Y)**.
4. Select all required nodes in sequence by moving mouse point to node and clicking.
   The **Select** button at the bottom left of the dialog can be used to select nodes automatically by the input minimum node number, maximum node number, and number increment.
   The **Clear** button at the right of the Select button can be used to clear all selected nodes.
5. Click the **Save** button to save the selected nodes and variable values to a file. The saved file can be imported into Microsoft Excel.
6. Click **Draw Graph** and then the graph can be drawn, as shown in Fig. 5.28.
   The circle marks in the graph are the original values of stress and the line is the curve approximated by the polynomial function.

7. Change the horizontal axis of the graph from node number to a coordinate of *X* or *Y* or *Z* and a coordinate versus stress graph can be drawn.

5.6 Q&A FOR SAMPLE DATA

Q1: What samples are saved on the book's website?

A1: To help readers to understand the basic concepts and behaviors of computational welding mechanics, 21 samples with six types of FEM geometry and computed results are given on the book's website. Details of the input files and output files are listed in Table 5.4.

Q2: What behaviors can be simulated using DATA1–DATA5?

A2: In DATA1-DATA4, the same temperature is applied to all nodes, and the maximum temperature in the welding thermal cycle is assumed to be 50°C, 150°C, 300°C, and 600°C, respectively. The thermal cycle with high temperature occurs at the location near the weld line. Using DATA1–DATA4, the thermal elastic stress-strain behavior and thermal elastic-plastic stress-strain behaviors at the heating and cooling processes can be simulated. To understand the creep behavior at high temperature, DATA5 was prepared in which one-dimensional bar is heated to 600°C and then kept for 10 hours to produce creep strain. The basic concepts in thermal elastic-plastic creep analysis and results have been described in Section 4.2. Here, mesh model, thermal conditions, and operation steps are explained for readers to practice the simulation as follows.

Mesh model:

- The length, width (height), and thickness of a one-dimensional bar are 200 mm, 20 mm, and 10 mm, respectively. The mesh division is shown in Fig. 5.29. The *x*-displacement at two ends is constrained.

Thermal conditions:

- In thermal elastic-plastic FEM analysis, nodal forces are produced by thermal expansion of material during heating and cooling cycles. Figure 5.30 shows the temperature changes in DATA1–DATA5. The points A, B, C, D, and E are the points of the maximum-reached temperature in thermal cycles.

Steps to start simulation:

- Q4 and A4 in Section 5.1 have described the general simulation step for using the FEM programs on the website. The following steps are for DATA1–DATA5:

    1. Copy **PROG\tepc2d.exe**, **DATA1\tepc2d.inp**, and **DATA1 \TEMP.FILE** to your data folder DATA1. Then double-click the program file **tepc2d.exe** to start computation.
    2. Copy **PROG\tepc2d.exe**, **DATA2\tepc2d.inp**, and **DATA2 \TEMP.FILE** to your data folder DATA2. Then double-click the program file **tepc2d.exe** to start computation.

TABLE 5.4 Sample Data on Book's Website

Example Types	Example Descriptions	Data Folders	Input Files	Result Files	Progams
One-dimensional constraint bar model	Maximum temperature is 50°C (thermal elastic behaviors)	CD:\DATA1	tepc2d.inp, TEMP.FILE	ehist.txt	**tepc2d.exe**
	Maximum temperature is 150°C (thermal elastic-plastic behaviors)	CD:\DATA2	tepc2d.inp TEMP.FILE	ehist.txt	
	Maximum temperature is 300°C (thermal elastic-plastic behaviors)	CD:\DATA3	tepc2d.inp TEMP.FILE	ehist.txt	
	Maximum temperature is 600°C (thermal elastic-plastic behaviors)	CD:\DATA4	tepc2d.inp EMP.FILE	ehist.txt	
	Heating up to 600°C and keeping for creep (thermal elastic-plastic creep behaviors)	CD:\DATA5	tepc2d.inp TEMP.FILE	ehist.txt	
Heat source	Moving welding heat source	CD:\DATA6	heat2d.inp	heat2d.post	**heat2d.exe**
	Fixed welding heat source	CD:\DATA7	heat2d.inp	heat2d.post	
Slit weld	Heat conduction using fixed welding heat source	CD:\DATA8	heat2d.inp	heat2d.post	**heat2d.exe**
	Thermal elastic-plastic analysis for thermal stress and residual stress	CD:\DATA9	tepc2d.inp TEMP.FILE	tepc2d.post	**tepc2d.exe**
	Verification of residual stress reproduced by inherent strain method	CD:\DATA10	inhs2d.inp	inhs2d.post	**inhs2d.exe**
Butt weld of thin plates	Heat conduction analysis by moving welding heat source	CD:\DATA11	heat2d.inp	heat2d.post	**heat2d.exe**

(Continued)

TABLE 5.4 Sample Data on Book's Website—cont'd

Example Types	Example Descriptions	Data Folders	Input Files	Result Files	Progams
	Thermal elastic-plastic analysis for deformation and residual stress	CD:\DATA12	tepc2d.inp TEMP.FILE	tepc2d.post	**tepc2d.exe**
	Verification of residual stress and deformation reproduced by inherent strain method	CD:\DATA13	inhs2d.inp	inhs2d.post	**inhs2d.exe**
	Residual stress measurement by inherent strain method	CD:\DATA14	inhs2d.inp	inhs2d.post	
	Prediction of residual stress using inherent strain database	CD:\DATA15	inhs2d.inp	inhs2d.post	
Thermal cutting of thin plate	Heat conduction analysis using moving welding heat source	CD:\DATA16	heat2d.inp	heat2d.post	**heat2d.exe**
	Thermal elastic-plastic analysis for thermal cutting deformation	CD:\DATA17	tepc2d.inp TEMP.FILE	tepc2d.post	**tepc2d.exe**
	Verification of thermal cutting deformation reproduced by inherent strain method	CD:\DATA18	inhs2d.inp	inhs2d.post	**inhs2d.exe**
Bead weld of thick plate	Heat conduction analysis using fixed heat source	CD:\DATA19	heat2d.inp	heat2d.post	**heat2d.exe**
	Thermal elastic-plastic analysis for angular deformation	CD:\DATA20	tepc2d.inp TEMP.FILE	tepc2d.post nhist.txt	**tepc2d.exe**
	Verification of angular deformation reproduced by inherent strain method	CD:\DATA21	inhs2d.inp	inhs2d.post	**inhs2d.exe**

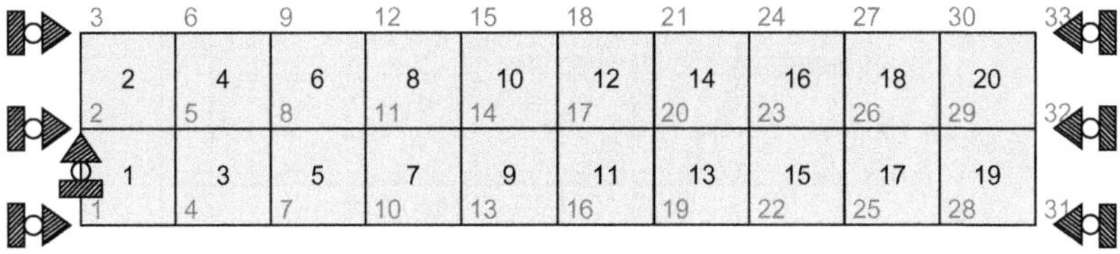

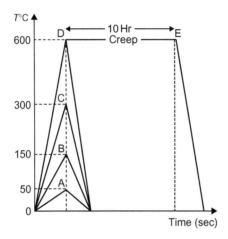

FIG. 5.29   Mesh division of one-dimensional constrained bar.

FIG. 5.30   Thermal cycles undergoing in a constrained bar.

3. Copy **PROG\tepc2d.exe**, **DATA3\tepc2d.inp**, and **DATA3 \TEMP.FILE** to your data folder DATA3. Then double-click the program file **tepc2d.exe** to start computation.

4. Copy **PROG\tepc2d.exe**, **DATA4\tepc2d.inp**, and **DATA4 \TEMP.FILE** to your data folder DATA4. Then double-click the program file **tepc2d.exe** to start computation.

5. Copy **PROG\tepc2d.exe**, **DATA5\tepc2d.inp**, and **DATA5\TEMP. FILE** to your data folder DATA5. Then double-click the program file **tepc2d.exe** to start computation.

Computed results:

- Figures 5.31 and 5.32 respectively show the stress changes in various thermal cycles and during creep process at 600°C, computed using DATA1–DATA5. The details have been described in Section 4.6.

Q3: What behaviors can be simulated by DATA6 and DATA7?

A3: A moving welding heat source model and a fixed welding heat source model have been described in Section 5.2. DATA6 and DATA7 are the input data for both heat source models, respectively. By performing heat

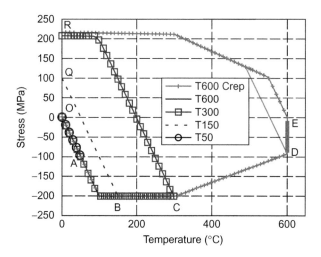

FIG. 5.31   Stress temperature curves of a restrained bar under a thermal cycle.

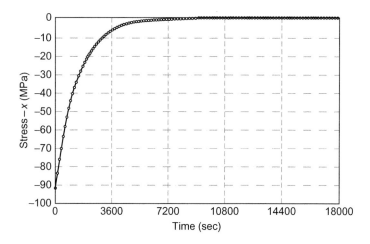

FIG. 5.32   Stress change due to creep at 600°C of holding temperature.

conduction simulation using a moving heat source DATA6 and a fixed heat source DATA7, the differences in results due to welding heat source models can be understood. The details of DATA6 and DATA7 are summarized as follows.

Mesh model:

- A square plate ($200 \times 200 \times 6$ mm) with a welding line in axis x is taken as a simulation model. Considering the geometrical symmetry, only half of the square plate is divided into elements, as shown in Fig. 5.33.

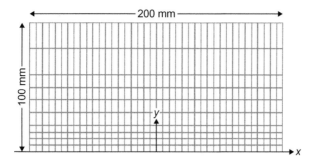

FIG. 5.33 Geometry sizes and element division of a half model.

Thermal conditions:

- DATA6 is a moving heat source data in which the welding speed is 3 mm/sec.
- DATA7 is a fixed heat source data in which the heating time is 1 sec.

Starting simulation:

- The Q4 and A4 in Section 5.1 have described the general simulation steps for using the FEM programs on the book's website. The following steps are for DATA6 and DATA7:

   1. Copy **PROG\heat2d.exe** and **DATA6\heat2d.inp** to your data folder DATA6. Then double-click the program file **heat2d.exe** to start computation.
   2. Copy **PROG\heat2d.exe** and **DATA7\heat2d.inp** to your data folder DATA7. Then double-click the program file **heat2d.exe** to start computation.

Computed results:

- Figures 5.34 and 5. 35 respectively show the maximum temperature distributions in the transverse direction and the welding line computed using DATA6 and DATA7. The maximum temperature near the weld line using a fixed heating source DATA7 with 1 sec heating time is higher than that using the moving heat source DATA6. The maximum temperatures away from the weld line using DATA6 and DATA7 are almost the same.

Q4: What behaviors can be simulated using DATA8–DATA10?

A4: DATA8 and DATA10 are the examples for slit welding. The thermal expansion of a slit weld is strongly constrained and large constraint stresses are induced. It is used to evaluate cold crack sensitivity under high constraint conditions. The constraint stress and plastic strains can be predicted using this data as input. DATA8 is the input data for heat conduction

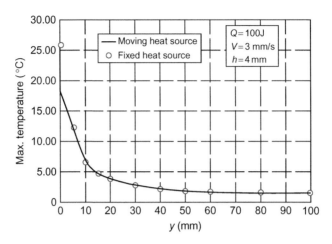

FIG. 5.34   Maximum-reached temperature in the thermal cycle in the transverse direction.

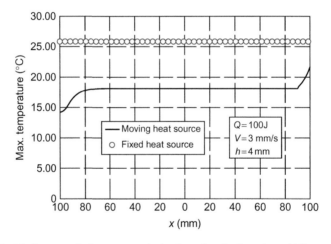

FIG. 5.35   Maximum-reached temperature in the thermal cycle along the weld line.

analysis and DATA9 is the input data for thermal elastic-plastic analysis. DATA10 is an example used to reproduce constraint stress in a slit weld using the inherent strain method. The details of DATA8–DATA10 are summarized as follows.

Mesh model:

- A slit weld is put on a square plate. The size of the plate is 400 mm in length, 400 mm in width, and 10 mm in thickness. The size of the slit weld is 100 mm in length, 10 mm in width, and 10 mm in thickness.

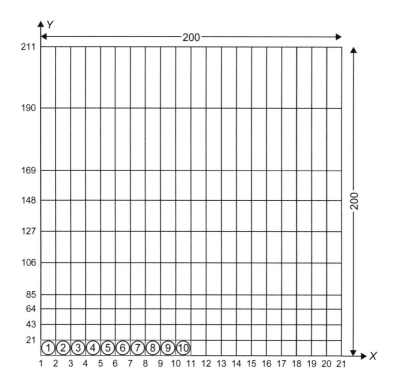

FIG. 5.36 Mesh model of a slit weld specimen.

Considering the geometric symmetry, only a quarter of the square plate is divided by the mesh, as shown in Fig. 5.36. The element number of the slit weld is from 1–10.

Thermal and mechanical boundary conditions:

- A fixed welding heat source with 1 sec heating time is adopted in the simulation. Thermal and mechanical symmetric conditions at axis x and axis y are included in the input data.

Steps to start simulation:

- The following steps are for DATA8–DATA10:

    1. Copy **PROG\heat2d.exe** and **DATA8\heat2d.inp** to your data folder DATA8. Then double-click the program file **inhs2d.exe** to start computation.
    2. Copy **PROG\tepc2d.exe** and **DATA9\tepc2d.inp** to your data folder DATA9. Then double-click the program file **tepc2d.exe** to start computation.

3. Copy **PROG\inhs2d.exe** and **DATA10\inhs2d.inp** to your data folder DATA10. Then double-click the program file **inhs2d.exe** to start computation.

Computed results using DATA8:

- Figure 5.37 shows the distribution of the maximum-reached temperature in the thermal cycle.
- Figure 5.38 shows temperature distributions along the weld line at different times and the maximum temperature reached in the thermal cycle.
- Figure 5.39 shows distributions of transient temperature in the transverse direction at different times and the maximum temperature reached in the thermal cycle.

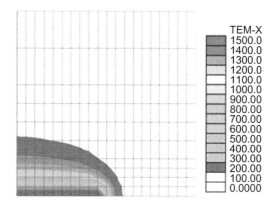

FIG. 5.37　Distribution of maximum-reached temperature in the thermal cycle.

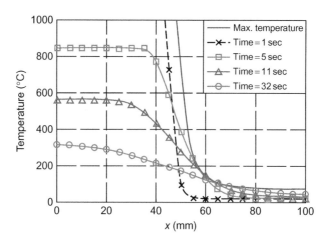

FIG. 5.38　Temperature along weld line.

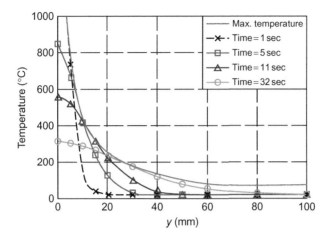

FIG. 5.39 Distributions of transient temperature in the transverse direction at different times and the maximum temperature reached in the thermal cycle.

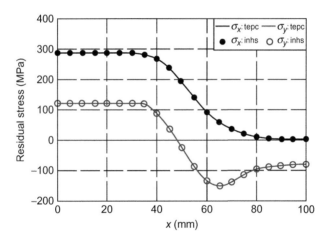

FIG. 5.40 Residual stresses σ_x and σ_y along the weld line.

Computed results using DATA9:

- Figure 5.40 shows distributions of residual stresses σ_x and σ_y along the weld line. Solid lines indicate the results by thermal elastic-plastic analysis (tepc) and circles by the inherent strain method (inhs). Both σ_x and σ_y are tensile in the weld line.
- Figure 5.41 shows distributions of residual stresses σ_x and σ_y in the middle transverse section ($x = 0$). Solid lines indicate the results by thermal elastic-plastic analysis and circles by the inherent strain method.

- Figure 5.42 shows distributions of residual plastic strains $(\varepsilon_x^p, \varepsilon_y^p, \gamma_{xy}^p)$ along the weld line. The results are obtained by thermal elastic-plastic analysis.
- Figure 5.43 shows distributions of residual plastic strains $(\varepsilon_x^p, \varepsilon_y^p, \gamma_{xy}^p)$ in the middle transverse section $(x = 0)$. The results are obtained by thermal elastic-plastic analysis.

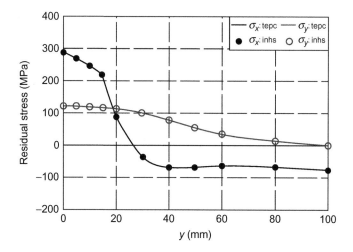

FIG. 5.41   Residual stresses σ_x and σ_y in the middle transverse direction.

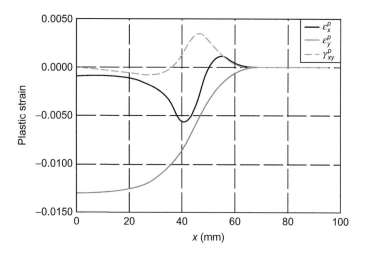

FIG. 5.42   Plastic strains $(\varepsilon_x^p, \varepsilon_y^p, \gamma_{xy}^p)$ along the weld line.

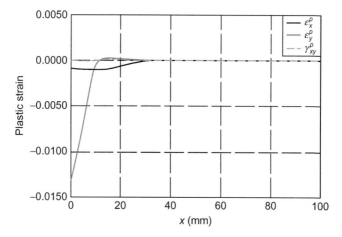

FIG. 5.43 Plastic strains $(\varepsilon_x^p, \varepsilon_y^p, \gamma_{xy}^p)$ in the transverse direction.

Computed results using DATA10:

- The circle marks (∘ •) shown in Figs 5.40 and 5.41 are the residual stresses (σ_x, σ_y) reproduced by the inherent strains shown in Figs 5.42 and 5.43.

Q5: What behaviors can be simulated using DATA11–DATA15?
A5: DATA11–DATA15 are the examples for a real size butt weld. The details, including data preparation, program running, and a discussion of the results, are given in Chapter 6.

Q6: What behaviors can be simulated using DATA16–DATA18?
A6: DATA16–DATA18 are examples for the simulation of heat conduction, thermal deformation, and residual stress due to thermal cutting using programs **heat2d.exe**, **tepc2d.exe**, and **inhs2d.exe**, respectively. The details for DATA16–DATA18 are summarized as follows.

Mesh model:

- It is assumed that a rectangle plate 1500 mm in length, 300 mm in width, and 6 mm in thickness is cut off. After thermal cutting, the cutting line will become free. Therefore, the shape and size after cutting is only considered in the mesh model shown in Fig. 5.44 by approximation.

Thermal and mechanical boundary conditions:

- Thermal cutting heat is applied to the elements at the cutting line. The cutting speed is 3 mm/sec.
- To prevent rigid movement of the plate, three points A, B, and C are fixed in deformation simulation.

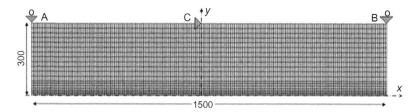

FIG. 5.44 Mesh division and boundary condition.

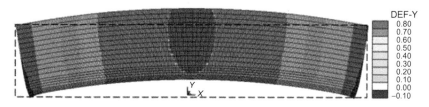

FIG. 5.45 Residual deformation due to thermal cutting using DATA17.

Steps to start simulation for DATA16–DATA18:

1. Copy **PROG\heat2d.exe** and **DATA16\heat2d.inp** to your data folder DATA16. Then double-click the program file **heat2d.exe** to start computation.
2. Copy **PROG\tepc2d.exe** and **DATA17\tepc2d.inp** to your data folder DATA17. Then double-click the program file **tepc2d.exe** to start computation.
3. Copy **PROG\inhs2d.exe** and **DATA18\inhs2d.inp** to your data folder DATA18. Then double-click the program file **inhs2d.exe** to start computation.

Computed results:

- Figure 5.45 shows the residual deformation due to thermal cutting using DATA17.
- Figure 5.46 indicates the displacement distribution along the cutting line (*x*-axis shown in Fig. 5.44). The solid line in this figure is computed by the program **tepc2d.exe** using DATA17, and the open circles O in the figure are computed by the program **inhs2d.exe** using DATA18.

Q7: What behaviors can be simulated using DATA19–DATA21?
A7: DATA19–DATA21 are examples for the simulation of heat conduction, angular deformation, and residual stress due to bead welding on the surface of a thick plate using programs **heat2d.exe**, **tepc2d.exe**, and **inhs2d.exe**, respectively. The details of DATA19–DATA21 are summarized as follows.

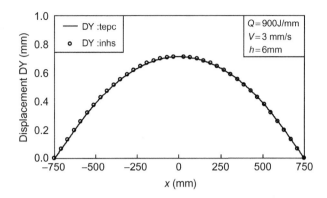

FIG. 5.46　Displacement in Y-direction, DY, along cutting line computed by **tepc2d.exe** (line) and **inhs2d.exe** circles.

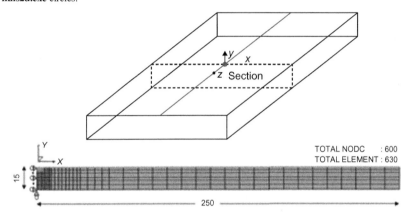

FIG. 5.47　A long bead-on-plate weld model and mesh model in simulation.

Mesh model:

- Figure 5.47 shows a long bead-on-plate weld model (500 mm in width, 15 mm in thickness) and mesh model in simulation. A half of the middle transverse section with a unit weld line is the model for analysis and it is divided into meshes considering the geometrical symmetry.

Thermal and mechanical boundary conditions:

- If it is assumed that welding current I, voltage U, welding speed V, and heat efficiency η are 120A, 15V, 3 mm/sec, and 0.5, respectively, the welding heat per unit weld line will be $Q = \eta IU/V = 300 (\text{J/mm})$. If the length of the weld line, width b, and depth h of the welding zone in the simulation model shown in Fig. 5.47 are 1 mm, 10 mm, and 3 mm, respectively, the heating time and heat generation rate per volume can be calculated based on Eq. (5.2.3), as described in Section 5.2.

Heating time:

- Welding time or heating time can be calculated using the unit welding line (1 mm) and the welding speed V (3 mm/sec):

$$t_w = 1.0/3.0 = 0.333 \text{ sec.}$$

- Heat generation rate per volume in welding metal:

$$\dot{q}_V = \frac{\eta \times I \times U}{V \times b \times h} \cdot \frac{1}{t_w} = \frac{0.5 \times 120 \times 15}{3 \times 10 \times 3} \cdot \frac{1}{0.333} = 30 \text{ J/(mm}^3 \cdot \text{sec)}.$$

- In heat conduction analysis, axis y is thermally insulated.
- In thermal deformation and stress analysis, nodes at axis y are fixed in the x direction.

Starting simulation for DATA19–DATA21:

1. Copy **PROG\heat2d.exe** and **DATA19\heat2d.inp** to your data folder DATA19. Then double-click the program file **heat2d.exe** to start computation.
2. Copy **PROG\tepc2d.exe** and **DATA20\tepc2d.inp** to your data folder DATA20. Then double-click the program file **tepc2d.exe** to start computation.
3. Copy **PROG\inhs2d.exe** and **DATA21\inhs2d.inp** to your data folder DATA21. Then double-click the program file **inhs2d.exe** to start computation.

Computed results:

- Figure 5.48 shows distribution of the maximum-reached temperature in thermal cycle using DATA19.
- Figure 5.49 shows the change of thermal deformation (y displacement at edge) with time during welding and cooling and residual angular distortion computed using DATA20.

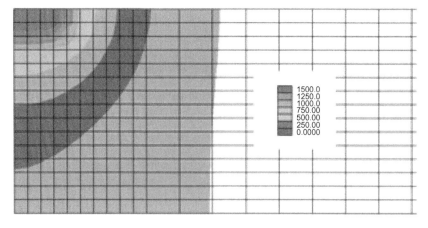

FIG. 5.48   Maximum temperature distribution computed using DATA19.

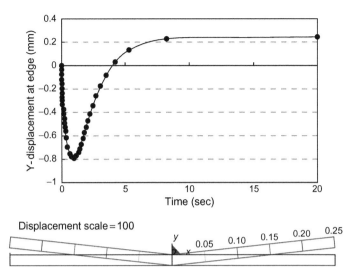

FIG. 5.49 History of edge deflection and residual angular distortion computed by **tepc2d.exe** using DATA20.

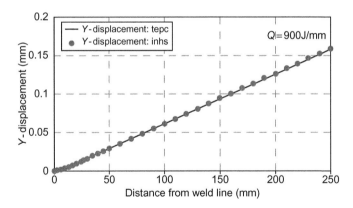

FIG. 5.50 Distributions of deflection in width direction computed by **tepc2d.exe** and **inhs2d.exe** using DATA20 and DATA21.

• Figure 5.50 shows the distribution of residual deflection (y displacement) along the width direction. The solid line in this figure is computed by the program **tepc2d.exe** using DATA20. The circle • in this figure are computed by the program **inhs2d.exe** using DATA21.

REFERENCE

[1] Goldak J, Chakravarti A, Bibby M. Computer modeling of heat flow in welds. Metallurgical Trans. 1984;17B:587–600.

Chapter 6

Simulation Procedures for Welding Heat Conduction, Welding Deformation, and Residual Stresses Using the FEM Programs Provided on the Companion Website

In Chapter 5, the following three executable FEM programs included on the book's companion website were briefly introduced:

- **heat2d.exe** for welding heat conduction analysis
- **tepc2d.exe** for thermal elastic-plastic creep analysis
- **inhs2d.exe** for simulation elastic analysis based on the inherent strain method

In this chapter, actual analyses on a butt-welded joint are performed using these three FEM programs. To demonstrate how to use these programs, we will explain, in detail, data preparation and setting conditions.

Try to answer the following questions after simulation has been performed to make sure you understand the procedure:

- What results did you obtain by welding heat conduction simulation?
- What results did you obtain by thermal elastic-plastic simulation?
- What results did you obtain by inherent strain simulation?
- How much computational time was taken for the simulation using each of the three programs?
- Do you understand the reason the deformation and residual stresses obtained by the thermal elastic-plastic program and the inherent strain program are almost the same?

6.1 SIMULATION STEPS USING THE WELDING HEAT CONDUCTION FEM PROGRAM

6.1.1 Purpose and Simulation Conditions

6.1.1.1 Purpose

- To learn the procedures for how to use the welding heat conduction FEM program **heat2d.exe** and to learn how to prepare input data.
- To understand simulation results such as transient temperature distribution, maximum temperature in thermal cycles, melting zone, heat affect zone, and mechanical melting zone.

6.1.1.2 Object

- The simulation object is the butt-welded joint shown in Fig. 6.1. Welding is conducted along the x coordinate and the y coordinate is the width direction of the plate.
- The dimensions of the butt-welded joint are 1500 mm in length, 600 mm in width, and 6 mm in thickness.

6.1.1.3 Welding Conditions

The following welding conditions [1] are used for this butt weld:

- welding current: $I = 200$ A
- welding voltage: $U = 18$ V
- welding speed: $v = 3$ mm/sec
- x coordinate of initial position of welding arc: $xini = -750$ mm
- x coordinate of end position of welding arc: $xend = +750$ mm
- width of weld bead: $b = 10$ mm

6.1.1.4 User Manual

The user manual is saved in **Manual\heat2d_manual.pdf**.

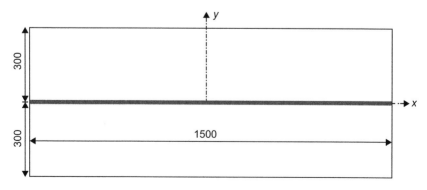

FIG. 6.1 Dimensions of butt-welded joint.

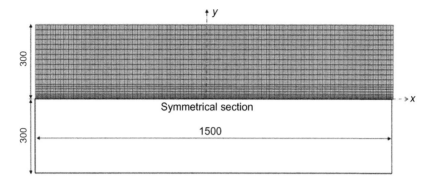

FIG. 6.2   Mesh division of butt-welded joint.

6.1.2 Preparation of Input File

Based on the manual, **Manual\head2d_manual.pdf**, 29 input cards (card0–card29) must be prepared as shown in the following and saved in the data file **heat2d.inp**.

6.1.2.1 Input Data for Mesh Division

Mesh Division

Since the sizes and welding heat conditions for two plates shown in Fig. 6.1 are the same, thermal conduction on the two sides of the weld line will be symmetrical. Considering the symmetry, only half of the butt-welded joint is modeled by FEM mesh as shown in Fig. 6.2.

Element Size

The minimum mesh size is 5 mm near the weld line and the maximum mesh size is 20 mm away from the weld line.

- total number of elements: NE = 6000
- total number of nodes: NP = 6321
- thickness of plate: thickness = 6 mm

List of Input Data

In the following lines, the characters and digital data with a gray background are copies of the input data for mesh division. Brief descriptions of the input data are given on the right side of each input line.

```
$Heat transfer FEM model of butt welding        ←Card0: title
6321                                             ←Card1: NP
1-750.0000000000 0.0000000000000 0.0000000000000 ←Card2: node
                                                   and x,y,z cord
```

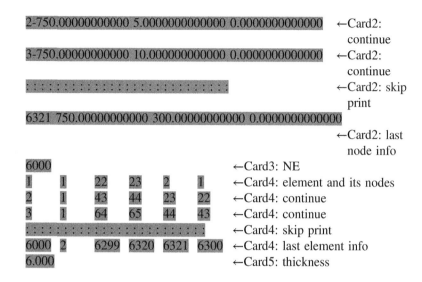

2-750.00000000000 5.0000000000000 0.0000000000000   ←Card2: continue

3-750.00000000000 10.000000000000 0.0000000000000   ←Card2: continue

⋮   ←Card2: skip print

6321 750.00000000000 300.00000000000 0.0000000000000
  ←Card2: last node info

6000   ←Card3: NE

1   1   22   23   2   1   ←Card4: element and its nodes

2   1   43   44   23   22   ←Card4: continue

3   1   64   65   44   43   ←Card4: continue

⋮   ←Card4: skip print

6000 2   6299 6320 6321 6300   ←Card4: last element info

6.000   ←Card5: thickness

6.1.2.2 Input Data for Material Properties

List of Material Properties

The material properties used in the heat conduction simulation are density ρ, specific heat c, thermal conductivity λ, and the equivalent transfer coefficient β for both heat convection and radiation.

Values of Material Properties

The values of the material properties with a change of temperature are shown in Table 6.1.

TABLE 6.1 Material Properties Used in Heat Conduction Analysis

$T°C$	$\rho[Kg/mm^3]$	$\lambda[J/(mm\ s°C)]$	$c[J/(Kg°C)]$	$\beta[J/(mm^2s°C)]$
0	7.82×10^{-6}	5.317×10^{-2}	4.083×10^{2}	0.917×10^{-5}
100	7.79×10^{-6}	5.275×10^{-2}	4.708×10^{2}	1.041×10^{-5}
200	7.77×10^{-6}	5.15×10^{-2}	5.333×10^{2}	1.416×10^{-5}
300	7.76×10^{-6}	4.941×10^{-2}	5.958×10^{2}	2.041×10^{-5}
400	7.72×10^{-6}	4.65×10^{-2}	6.583×10^{2}	2.916×10^{-5}
500	7.69×10^{-6}	4.275×10^{-2}	7.208×10^{2}	4.041×10^{-5}
600	7.66×10^{-6}	3.817×10^{-2}	7.833×10^{2}	5.416×10^{-5}
700	7.64×10^{-6}	3.275×10^{-2}	8.458×10^{2}	7.041×10^{-5}
800	7.61×10^{-6}	2.65×10^{-2}	9.083×10^{2}	8.916×10^{-5}

List of Input Data

In the following lines, the characters and digital data with a gray background are copies of the input data for the material properties. Brief descriptions of the input data are given on the right side of each input line.

9	←Card6: NDENS is temperature points for ρ
0.00000, 7.82e-06	←Card7: $T°C$, ρ
100.000, 7.79e-06	←Card7: continue
200.000, 7.77e-06	←Card7: continue
300.000, 7.74e-06	←Card7: continue
400.000, 7.72e-06	←Card7: continue
500.000, 7.69e-06	←Card7: continue
600.000, 7.66e-06	←Card7: continue
700.000, 7.64e-06	←Card7: continue
800.000, 7.61e-06	←Card7: end at the NDENS-th point
9	←Card8: NLAMD is temperature points for λ
0.0000000, 5.31667E-02	←Card9: $T°C$, λ
100.00000, 5.27500E-02	←Card9: continue
200.00000, 5.15000E-02	←Card9: continue
300.00000, 4.94167E-02	←Card9: continue
400.00000, 4.65000E-02	←Card9: continue
500.00000, 4.27500E-02	←Card9: continue
600.00000, 3.81667E-02	←Card9: continue
700.00000, 3.27500E-02	←Card9: continue
800.00000, 2.65000E-02	←Card9: end at the NLAMD-th point
9	←Card10: NCAPA is temperature points for c
0.0000000, 4.083333E+02	←Card11: $T°C$, c
100.00000, 4.708333E+02	←Card11: continue
200.00000, 5.333333E+02	←Card11: continue
300.00000, 5.958333E+02	←Card11: continue
400.00000, 6.583333E+02	←Card11: continue
500.00000, 7.208333E+02	←Card11: continue
600.00000, 7.833333E+02	←Card11: continue
700.00000, 8.458333E+02	←Card11: continue
800.00000, 9.083333E+02	←Card11: end at the NCAPA-th point
9	←Card12: NBETA is temperature points for β
0.0000000, 9.166667E-06	←Card13: $T°C$, β
100.00000, 1.041667E-05	←Card13: continue
200.00000, 1.416667E-05	←Card13: continue
300.00000, 2.041667E-05	←Card13: continue
400.00000, 2.916667E-05	←Card13: continue
500.00000, 4.041667E-05	←Card13: continue
600.00000, 5.416667E-05	←Card13: continue
700.00000, 7.041667E-05	←Card13: continue
800.00000, 8.916667E-05	←Card13: end at the NBETA-th point

6.1.2.3 Input Data for Thermal Boundary Conditions

- Environmental temperature: TROOM = 20°C.
- Number of initial temperatures at specified nodes: NINIT = 0.
  If NINIT = 0, the initial temperatures for all nodes are automatically set to TROOM.
- Number of fixed temperature at specified nodes: NFIX = 0.
  If NFIX = 0, the fixed temperature does not need to be defined.
- Setting of element surfaces for heat transfer surface: NFACE(1–6).
- Symmetric plane along the welding line is an insulation surface: NFACE(1) = 0.
- Edge surfaces through the plate thickness are assumed to be insulation surfaces because the surface area is very small: NFACE(2) = NFACE(2) = NFACE(3) = NFACE(4) = 0.
- Top surface and low surface of plate are heat transfer surface: NFACE (5) = NFACE (6) = 1.

List of Input Data

In the following lines, the characters and digital data with a gray background are copies of the input data for the thermal boundary conditions. Brief descriptions of the input data are given on the right side of each input line.

```
20.000,              ←Card14: TROOM
0.                   ←Card15: NINIT (if NINIT = 0, skip card16)
0.                   ←Card17: NFIX (if NFIX = 0, skip card18)
0, 0, 0, 0, 1, 1,    ←Card19: NFACE(1)–NFACE(6)
```

6.1.2.4 Input Data for Welding Heat

- Heating time or welding time: TWELD = 500 sec.
- The welding time is determined by dividing the weld length L by the welding speed v, which are 1500 mm and 3 mm/sec, respectively.
- The heat source parameters are a, b, and h.
- It is assumed that welding heat is distributed uniformly in a simple rectangular prism. The sizes of the rectangular prism are $a = 10$ mm in length, $b = 10$ mm in width, and $h = 6$ mm in depth.
- The volume heat generation rate is $\dot{q}_V = 4.5$ [J/(mm³sec)].
- The $\dot{q}_V$ is an input parameter for the simplified moving heat source shown in Fig. 6.3, which is determined by Eq. (6.1.1) using welding conditions (current I, voltage U) and effective heat efficiency η:

$$\dot{q}_V = \eta \frac{IU}{abh} = 0.75 \frac{200 \times 18}{10 \times 10 \times 6} = 4.5 [\text{J/(mm}^3\text{sec)}]. \qquad (6.1.1)$$

- The total number of elements for volume heat generation is NEQV = 300.
- The element number of the heat generation elements is from 1 to 300 on the weld line.

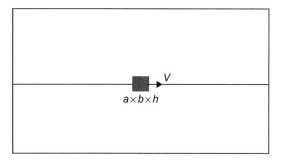

FIG. 6.3 Simplified moving heat source for butt welding.

- The starting time and ending time for each heat generation element is TWS = 0 and TWE = 0.
- TWS = 0 and TWE = 0 mean that the starting time and ending time are automatically determined by the program.
- The total number of element surfaces exposed to heat flux is NEQF = 0.
- If NEQF = 0, the surface heat flux does not need to be defined.
- The flag for the moving heat source or fixed heat source is MV = 1.
- MV = 1 means the moving heating source is supplied on welding.
- The x coordinate of the initial position of the welding heat source is XINI = −750 mm.
- The welding speed is $V = 3$ mm/sec.

List of Input Data

In the following lines, the characters and digital data with a gray background are copies of the input data relating to the welding heat source. Brief descriptions of the input data are given on the right side of each input line.

500.0,	←Card20: WTIME
300,	←Card21: NEQV
1 4.500, 0, 0,	←Card22: IE,QV,TWS,TSE
2 4.500, 0, 0,	←Card22:: continue
3 4.500, 0, 0,	←Card22: continue
: : : : : : : : : :	←Card22: print skipped
299 4.500, 0, 0,	←Card22: continue
300 4.500, 0, 0,	←Card22: continue
0	←Card23: NEQF (If NEQF = 0, skip card24)
1	←Card25: MV
3.00, −750.0, 10.00, 5.00,	←Card26: V, XINI, AMX, AMY

6.1.2.5 Control Parameters of Computation

- The time to the end of heat conduction analysis is TIME = 72,000 sec.
- The initial time step for heat conduction computation is DT0 = 0.01 sec.

- The steps to the end of computation are NSTEP = 100,000.
- The output interval steps of the nodal temperature file **TEMP.FILE** are NOUT = 50.
- The output interval steps of the postprocessing file **heat2d.post** are MOUT = 500.
- The temperature range for the average cooling rate is $TC1 = 800°C$, $TC2 = 500°C$.

List of Input Data

In the following lines, the characters and digital data with a gray background are copies of the input data relating to the control parameters of computation. Brief descriptions of the input data are given on the right side of each input line.

7200.0, 0.01,	←Card27: TIME, DT0
100000, 50, 500,	←Card28: NSTEP, NOUT, MOUT
800.00, 500.00,	←Card29: $C1$, $TC2$

6.1.2.6 Save Input Data

After the input data is prepared, save the input file **heat2d.inp** in the computation folder as follows: **DATA11\heat2d.inp**.

6.1.3 Steps to Execute heat2d.exe for Welding Heat Conduction

This procedure can be performed as follows:

1. Create a computation folder called **DATA11**.
2. Copy the prepared input data from **DATA11\heat2d.inp** to **DATA11**.
3. Copy the executable program to **DATA11**.
4. Double-click the program **heat2d.exe** to start the computation.

6.1.4 Viewing Results Using Postprocessing Program

6.1.4.1 Temperature Distribution During Welding

If the result file **heat2d.post** is read to the postprocessing program **awsd.exe**, the temperature distribution during welding, as shown in Fig. 6.4, can be viewed.

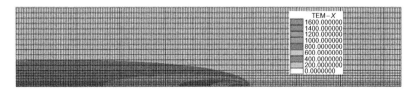

FIG. 6.4 Temperature distribution during welding.

The shape of the isothermal curves in the rear region of the welding arc is a long and narrow elliptical shape. The temperature gradient in the rear region of the welding arc is much smaller than that in the forward region.

6.1.4.2 Distribution of Maximum Temperature in the Thermal Cycle

By a similar operation, the distribution of the maximum temperature in the thermal cycle, as shown in Fig. 6.5, can be drawn by the postprocessing program **awds.exe**. The maximum temperature has a large gradient in the width direction of the plate and distributes almost uniformly in the welding direction.

6.1.4.3 Temperature Distribution Graph on the Weld Line and Transverse Section

Figures 6.6(a) and (b) show the temperature distributions on the welding line and the middle transverse section, respectively. **T:300 sec** and **T:max** in Figs 6.6(a) and (b) are the transient temperature during welding (time = 300 sec) and the maximum temperature in the thermal cycle, respectively. The maximum temperature has uniform distribution in the welding direction except near the edges of the weld line. The maximum temperature shown in Fig. 6.6(a) is lowest at the starting edge and highest at the ending edge. The heat transfer to the environment at the ending edge of the weld line is very small, and temperature increases are higher due to the preheating effect before the welding arc reaches the ending edge.

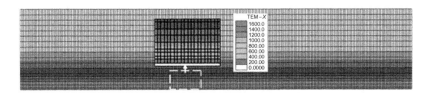

FIG. 6.5   Distribution of maximum temperature in welding thermal cycle.

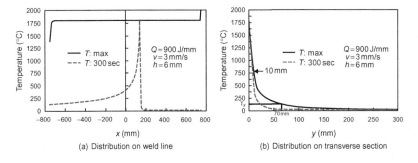

(a) Distribution on weld line   (b) Distribution on transverse section

FIG. 6.6   Temperature distributions on the welding line and in the middle.

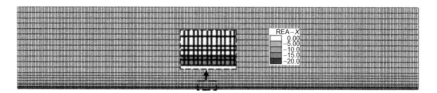

FIG. 6.7   Distribution of average cooling rate between 800°C and 500°C.

This phenomenon is called **thermal reflection**. Inversely, as the preheating does not exit at the starting edge of the welding line, the maximum temperature becomes lower.

From the transient temperature distribution at time = 300 sec shown in Figs 6.6(a) and (b), the length and a half width of the mechanical melting zone (775°C) during welding can be predicted, which are 55 mm and 10 mm, respectively.

From the maximum temperature distribution, a half width of the welded zone, where the temperature reaches the yielding temperature T_Y defined by Eq. (1.2.5) in Chapter 1, can be predicted and its size is about 70 mm, as shown in Fig. 6.6(b).

6.1.4.4 Average Cooling Rate Between 800°C and 500°C

Figure 6.7 shows the distribution of the average cooling rate between 800°C and 500°C. The average cooling rate on the weld line is about −17.5°C. The average cooling rate at the region in which the maximum temperature is below 800°C is not computed and the zero value is given, which is represented by white in Fig. 6.7.

6.2 SIMULATION STEPS USING THE THERMAL ELASTIC-PLASTIC CREEP FEM PROGRAM

6.2.1 Purpose and Simulation Conditions

6.2.1.1 Purpose

- To learn the detailed steps for using the welding thermal elastic-plastic creep FEM program **tepc2d.exe** and to learn how to prepare input data.
- To understand the simulation results, including thermal deformation, thermal stresses and residual deformation, residual stresses, and residual plastic strains.

6.2.1.2 Object

The simulation object is a butt-welded joint, as shown in Fig. 6.8. The dimensions of the butt-welded joint are the same as those in the welding heat conduction simulation described in Section 6.1.1.

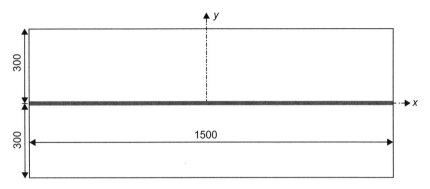

FIG. 6.8 Dimensions of butt-welded joint.

6.2.1.3 Welding Conditions

The following welding conditions [1] are used for this butt welding:

- welding current: $I = 200$ A
- welding voltage: $U = 18$ V
- welding speed: $v = 3$ mm/sec

6.2.1.4 User Manual

The user manual is saved in **Manual\tepc2d_manual.pdf**.

6.2.2 Preparation of Input File

Based on the manual, **Manual\tepc2d_manual.pdf**, 37 input cards (card0–card36) must be prepared as shown in the following and saved in the data file **tepc2d.inp**.

6.2.2.1 Input Data for Mesh Division

Mesh Division

The mesh division used for the thermal elastic-plastic creep simulation must be the same as the mesh used in the welding heat conduction simulation.

Element Size

The minimum mesh size is 5 mm near the weld line and the maximum mesh size is 20 mm away from the weld line.

- total number of elements: NE = 6000
- total number of nodes: NP = 6321
- thickness of plate: thickness = 6 mm

List of Input Data

In the following lines, the characters and digital data with a gray background are copies of the input data relating to the mesh division. Brief descriptions of the input data are given on the right side of each input line.

$Thermal elastic-plastic creep FEM model of butt welding	
	←Card0: TITLE
6321	←Card1: NP
1-750.00000000000 0.0000000000000 0.0000000000000	←Card2: node and cord
2-750.00000000000 5.0000000000000 0.0000000000000	←Card2: continue
3-750.00000000000 10.000000000000 0.0000000000000	←Card2: continue
::::::::::::::::::::::::::::	←Card2: skip print
6321 750.00000000000 300.00000000000 0.0000000000000	←Card2: last node info
6000	←Card3: NE
1 1 22 23 2 1	←Card4: element and its nodes
2 1 43 44 23 22	←Card4: continue
3 1 64 65 44 43	←Card4: continue
::::::::::::::::::::::::::::	←Card4: skip print
6000 2 6299 6320 6321 6300	←Card4: last elem. info
6.000	←Card5: thickness

6.2.2.2 Input Data for Material Properties

List of Material Properties

The material properties used in this simulation are the thermal expansion coefficient α, Young's modulus E, Poisson's ratio ν, yield stress σ_Y, and linear plastic work hardening coefficient H'.

Values of Material Properties

Table 6.2 shows the values of the material properties considering the changes in temperature.

The material properties at high temperature are difficult to prepare for simulation. The values used here are taken from published papers given in Appendix B on the companion website. Young's modulus E, yield stress σ_Y, and linear plastic work hardening coefficient H' at the temperature above the mechanical melting point are assumed to be about 1–10% of the values at room temperature to keep computation stable. Here, for yield stress σ_Y and plastic hardening coefficient H', 10% of these values at room temperature are used.

TABLE 6.2 Material Properties Used in Thermal Elastic-Plastic Simulation

$T°C$	E(MPa)	ν	σ_Y(MPa)	H'(MPa)	α(1/°C)
0	2×10^5	0.3	200	2×10^3	1×10^{-5}
300	2×10^5	0.3	200	2×10^3	1×10^{-5}
775	2×10^5	0.3	20	0.2×10^3	1×10^{-5}
1000	2×10^5	0.3	20	0.2×10^3	1×10^{-5}
1500	2×10^5	0.3	20	0.2×10^3	1×10^{-5}

- The creep strain flag and creep strain coefficients are NCREP = 0 and ACREP = 0.
- NCREP = 0 means the creep strain is not considered.
- ACREP = 0 means the creep strain rate is zero.
- NTCRP = 4 means four sets of temperature and scale factor of creep strain will be defined based on the function $f(T)$ given in Table 4.3 in Section 4.6.4.
- NSCRP = 4 means four sets of equivalent stress and scale factor of creep strain will be defined based on the function $g(s)$ given in Table 4.3 in Section 4.6.2.

List of Input Data

In the following lines, the characters and digital data with a gray background are copies of the input data relating to the material properties. Brief descriptions of the input data are given on the right side of each input line.

5	←Card6: NYMOD
0.0000000000000 200000.0000000	←Card7: $T1,E1$
300.00000000000 200000.0000000	←Card7: $T2,E2$
775.00000000000 20000.00000000	←Card7: $T3,E3$
1000.000000000 20000.00000000	←Card7: $T4,E4$
1500.000000000 20000.00000000	←Card7: $T5,E5$
2	←Card8: NPR
0.0000000000000 0.3000000000000	←Card9: $T1,\nu1$
15000.000000000 0.3000000000000	←Card9: $T2,\nu2$
5	←Card10: NYIELD
0.0000000000000 200.00000000000	←Card11: $T1,SY1$
300.00000000000 200.00000000000	←Card11: $T2,SY2$
775.00000000000 20.000000000000	←Card11: $T3,SY3$
1000.000000000 20.000000000000	←Card11: $T4,SY4$
1500.000000000 20.000000000000	←Card11: $T5,SY5$

```
5                                         ←Card12: NHARD
0.0000000000000 2000.0000000000          ←Card13: T1,H1′
300.00000000000 2000.0000000000          ←Card13: T2,H2′
775.00000000000 200.00000000000          ←Card13: T3,H3′
1000.0000000000 200.00000000000          ←Card13: T5,H4′
1500.0000000000 200.00000000000          ←Card13: T5,H5′
5                                         ←Card14: NALPH
0.0000000000000 1.00e-05                  ←Card15: T1,α1
300.00000000000 1.00e-05                  ←Card15: T2,α2
775.00000000000 1.00e-05                  ←Card15: T3,α3
1000.0000000000 1.00e-05                  ←Card15: T4,α4
1500.0000000000 1.00e-05                  ←Card15: T5,α5
0 0.000000                                ←Card16: NCREEP (creep flag),
                                            ACREEP (creep rate coef.)
4                                         ←Card17: NTCRP is the points of
                                            (T, creep coef.)
0.0000000000000 0.0000000000000          ←Card18: T1,f1
300.00000000000 0.0000000000000          ←Card18: T2,f2
600.00000000000 1.0000000000000          ←Card18: T3,f3
1000.0000000000 1.0000000000000          ←Card18: T4,f4
4                                         ←Card19: NSCRP is the points of
                                            (stress, creep coef.)
0.0000000000000 0.0000000000000          ←Card20: s1,g1
50.000000000000 0.5000000000000          ←Card20: s2,g2
100.00000000000 1.0000000000000          ←Card20: s3,g3
200.00000000000 1.0000000000000          ←Card20: s4,g4
```

6.2.2.3 Input Data for Welding Dummy Elements

- The total number of welding dummy elements is NWELD = 0.
- If NWELD > 0, the elements on the weld line prior to welding are just the dummy elements that have no stiffness and no stresses will be produced in them. After welding, the dummy elements are switched to normal elements during welding when they are heated by the welding heat source.
- If NWELD = 0, the elements on the weld line exist before and after welding. In this example, to understand the basic behaviors of welding deformation and stresses, it is assumed that many tack welds on the weld line exist before butt welding and then the welding dummy elements are not adopted here.

List of Input Data

In the following lines, the characters and digital data with a gray background are copies of the input data relating to the welding dummy elements. Brief descriptions of the input data are given on the right side of each input line.

0, 20000.0, 20.0, 0.0,   ←Card21: NWELD, WYM, WSY,WHD
※if NWELD = 0, card22 is not needed.

6.2.2.4 Input Data for Boundary Conditions

Symmetric Displacement Boundary

- The x-axis is a symmetric axis in this FEM model. Therefore, displacements in the y direction at nodes along the x-axis (301 nodes in total) are zero. To prevent rigid body movement in the welding direction, the displacement in the x direction at one node of the middle transverse section, as shown in Fig. 6.9, is constrained.
- The total number of constrained nodal displacements is NFIX = 302.
- The total number of nodal forces is NFORCE = 0.
- If NFORCE = 0, the equivalent nodal forces are not necessary to define in input data.

List of Input Data

In the following lines, the characters and digital data with a gray background are copies of the input data relating to boundary conditions. Brief descriptions on the input data are given on the right side of each input line.

302   ←Card23: NFIX is total number for disp. constraint
3151, 1, 0.   ←Card24: KFIXP, KFIXD, FIXDIS
1, 2, 0.   ←Card24: continue
22, 2, 0.   ←Card24: continue
: : : : : :   ←Card24: print skipped
6301, 2, 0.   ←Card24: end at NFIX-th disp. constraint
0   ←Card25: NFORCE is total number for nodal force
※if NFORCE = 0, card26 is not needed.

6.2.2.5 Input Data for Control Parameters of Computation

- The total steps of the nodal temperature file are NSTEP = 1002.
- NSTEP must be the same as the steps of temperature file **TEMP.FILE**.
- The convergent clearance is conv. = 0.0001.
- The stress and strain zero reset flag is NOSET = 0.

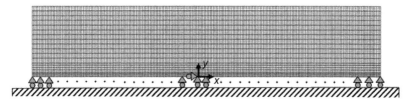

FIG. 6.9 Boundary conditions in welding deformation and stress simulation.

If N0SET = 0, the general elastic-plastic creep analysis is performed in computation. If N0SET > 0, the stresses and plastic strains are set to zero when the temperature is higher than the mechanical melting point.

- The environment temperature is TROOM = 20°C.
- The mechanical melting temperature is TWHAZ = 775°C.
- The maximum temperature cut is TCUT = 775°C.

Here, TCUT is used to save computational time and control the error of stress-strain computation at high temperature. If the temperature is higher than TCUT, the thermal strain due to expansion does not increase. Usually, TCUT is equal to or higher than the mechanical melting temperature and lower than the material melting temperature.

- The maximum temperature increment is dtempset = 5°C.
- The maximum time increment is dtimeset = 72,000 sec.

The "dtimeset" is only used in creep strain computation. If creep strain is ignored in computation, it is recommended that a large value be set.

- The output interval steps are istep_out = 1000.
- Istep_out is used to control the size of the output files **tepc2d.post, nhist.txt**, and **ehist.txt**, which are for postprocessor, node history variables, and element variables, respectively.
- The total node number for the node history variable output file is **nhist.txt**: np_out = 0.
  np_out = 0 means that the results are not written in **nhist.txt**.
- The total element number for the node history variable output file is **ehist. txt**: ne_out = 0.
  ne_out = 0 means that results are not written in **ehist.txt**.

List of Input Data

In the following lines, the characters and digital data with a gray background are copies of the input data concerning the computational control parameters. Brief descriptions of the input data are given on the right side of each input line.

```
1002,                              ←Card27: NSTEP
0.0001,                            ←Card28: CONV
0,                                 ←Card29: N0SET
20.00, 775.00, 775.000,           ←Card30: TROOM, TWHAZ, TCUT
5.00, 72000.0,                     ←Card31: dtempset, dtimeset
10000,                             ←Card32: istep_out
0,                                 ←Card33: np_out
if np_out = 0, card34 is not needed.
0,                                 ←Card35: ne_out
if ne_out = 0, card36 is not needed.
```

6.2.2.6 Save Input Data

After the input data is prepared, save the input file **tepc2d.inp** in the computation folder as follows: **DATA12\tepc2d.inp**.

6.2.3 Prepare the Node Temperature File

Besides the input file **tepc2d.inp**, the file **TEMP.FILE** for nodal temperature history, which has been already computed by **heat2d.exe**, needs to be copied to the computation folder from the heat conduction computation folder.

6.2.4 Steps to Execute the FEM Program

This procedure can be performed using the following steps:

1. Create a computation folder called **DATA12**.
2. Copy the prepared input data from **DATA11\tepc2d.inp** to **DATA12**.
3. Copy the program to **DATA12**.
4. Double-click the program **tepc2d.exe** to start the computation.

6.2.5 Viewing Results Using the Postprocessing Program

6.2.5.1 Distribution of Transient Thermal Stresses

If the result file **tepc2d.post** is read to the postprocessing program **awsd.exe**, the distribution of the transient thermal stress components σ_x and σ_y, as shown in Fig. 6.10, can be viewed. Longitudinal stress component σ_x in the welding direction exists in a

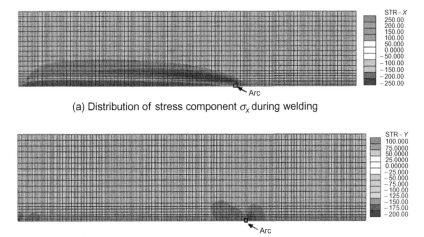

(a) Distribution of stress component σ_x during welding

(b) Distribution of transverse stress component σ_y during welding

FIG. 6.10 Distribution of transient thermal stresses during welding.

wide area behind the welding arc. The transverse stress component σ_y in the width direction distributes in a very narrow area around the welding arc and the welded left edge of the butt joint.

6.2.5.2 Distribution of Residual Stresses

Figures 6.11(a) and (b) represent the overall distribution of welding residual stresses σ_x and σ_y drawn using the postprocessing program. Longitudinal residual stress component σ_x shows almost uniform distribution along the welding line except near the two ends of the butt joint. The transverse residual stress component σ_y exists only near the two ends of the butt joint.

Figure 6.12(a) shows the distribution of welding residual stresses along the weld line. Longitudinal residual stress σ_x exhibits a large tensile value along the weld line and a zero value at the two ends of the butt joint. The transverse residual stress σ_y shows a large compressive value near the two ends of the weld line and becomes a small tensile value away from the two ends.

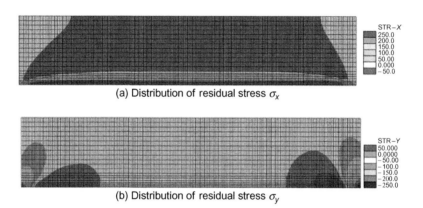

(a) Distribution of residual stress σ_x

(b) Distribution of residual stress σ_y

FIG. 6.11   Distribution of welding residual stresses.

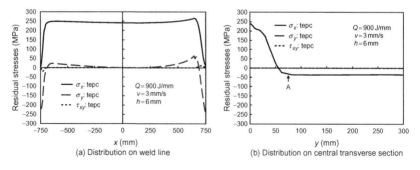

(a) Distribution on weld line

(b) Distribution on central transverse section

FIG. 6.12   Residual stress distributions along sections.

Figure 6.12(b) shows the distribution of welding residual stresses on the middle transverse section. Longitudinal residual stress σ_x indicates a tensile value in the vicinity of the welded zone and a compressive value away from the welded zone. The transition point A at the compressive side of the residual stress is the boundary between the plastic deformation zone and the elastic deformation zone during welding. The transverse stress σ_y and shear stress τ_{xy} are very small.

6.2.5.3 Welding Shrinkage Deformation

Figure 6.13 shows the shrinkage deformation in both the welding direction and transverse direction. Figures 6.14(a) and (b) show the distribution of longitudinal shrinkage ΔL in the transverse direction and transverse shrinkage ΔS in welding direction, respectively. The longitudinal shrinkage is larger near the welded zone and becomes small with the increase of transverse distance from the welded zone. The transverse shrinkage has almost a uniform value along the welding direction except at the two ends.

6.2.5.4 Distribution of Residual Plastic Strain

Figures 6.15(a) and (b) show the distribution of both the residual plastic strains ε_x^p and ε_y^p in the welding direction (x) and transverse direction (y), respectively.

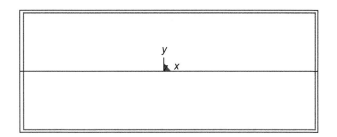

FIG. 6.13 Shrinkage deformation of butt joint (scaled 50 times).

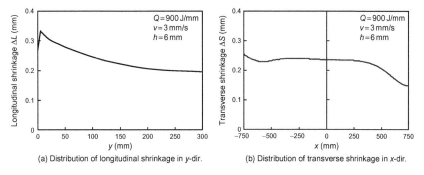

(a) Distribution of longitudinal shrinkage in y-dir.

(b) Distribution of transverse shrinkage in x-dir.

FIG. 6.14 Distribution of longitudinal shrinkage ΔL and transverse shrinkage ΔS.

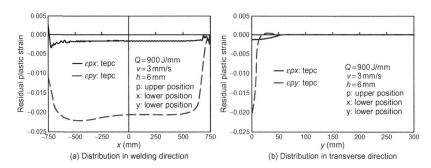

FIG. 6.15   Distribution of residual plastic strains in both the welding and transverse directions.

The residual plastic strains ε_x^p and ε_y^p have almost uniform distribution in the welding direction except the zones near the two ends of the weld line. The transverse plastic strain ε_y^p is about 10 times larger than the longitudinal plastic strain ε_x^p. This is why the transverse shrinkage is much larger than the longitudinal shrinkage and the transverse residual stress σ_y^p is much smaller than the longitudinal residual stress σ_x^p.

From the distribution of residual plastic strains ε_x^p and ε_y^p shown in Fig. 6.15(b), it can be seen that the residual plastic strains exist in the width of 55 mm from the weld line. The distribution shape of residual plastic strain ε_x^p in the transverse direction (y) is a simple trapezoidal type. Compared to the residual plastic strain ε_x^p, the residual plastic strain ε_y^p is quite a bit larger near the welded zone. However, this indicates a small tensile value in the zone between $y = 30$ mm and $y = 55$ mm from the welded zone and this is because the shrinkage near the welded zone is constrained by the metal slightly away from the welded zone.

6.3 SIMULATION STEPS USING THE INHERENT STRAIN FEM PROGRAM

6.3.1 Purpose and Simulation Conditions

6.3.1.1 Purpose

- To learn the detailed steps for how to use the inherent strain FEM program **inhs2d.exe** and to learn how to prepare the input data.
- To understand the simplicity of the inherent strain method and to confirm its accuracy of the reproduction of welding residual deformation and residual stresses.

6.3.1.2 Object

The analysis object is a butt-welded joint, as shown in Fig. 6.16. The sizes and mesh division of the butt-welded joint are the same as those shown in Fig. 6.2.

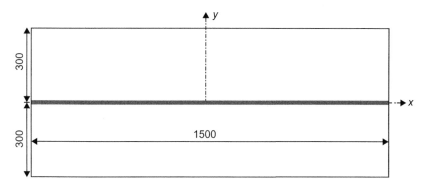

FIG. 6.16 Dimensions of butt-welded joint.

6.3.1.3 Conditions

The inherent strains are the input data in the simulation described in this section. The residual plastic strains computed by the thermal elastic-plastic FEM program in Section 6.2 are taken as the inherent strains for the input for the program **inhs2d.exe**.

6.3.1.4 User Manual

The user manual is saved in **Manual\inhs2d_manual.pdf**.

6.3.2 Preparation of Input File

Based on the manual, **Manual\inhs2d_manual.pdf**, the following input data (card0–card12) must be prepared and saved in the data file **inhs2d.inp**.

6.3.2.1 Input Data for Mesh Division

Mesh Division

The mesh shown in Fig. 6.17 is the same as the mesh used in the thermal elastic-plastic FEM program because plastic strains in the FEM elements will be used as the input data for the inherent strain program **inhs2d.exe**.

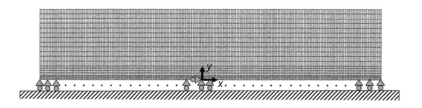

FIG. 6.17 Mesh division.

List of Input Data

In the following lines, the characters and digital data with a gray background are copies of the input data relating to mesh division. Brief descriptions of the input data are given on the right side of each input line.

```
$Inherent strain method model of butt welding    ←Card0: comments
6321                                              ←Card1: total number of nodes
1-750.00000000000 0.0000000000000 0.0000000000000 ←Card2: node ID
                                                       and its cord.
2-750.00000000000 5.0000000000000 0.0000000000000 ←Card2:
                                                       continue
3-750.00000000000 10.000000000000 0.0000000000000 ←Card2:
                                                       continue
:::::::::::::::::::::::::::::::::                 ←Card2: print skipped
6321 750.00000000000 300.00000000000 0.0000000000000  ←Card2: the
                                                       last node
6000                                             ←Card3: total number of elements
1 1 22 23 2 1                                    ←Card4: element ID and its nodes
2 1 43 44 23 22                                  ←Card4: continue
3 1 64 65 44 43                                  ←Card4: continue
:::::::::::::::::::::::::::::::::                 ←Card4: print skipped
6000 2 6299 6320 6321 6300                       ←Card4: the last element
6.000                                            ←Card5: thickness
```

6.3.2.2 Input Data for Material Properties

Material Properties

Young's modulus and Poisson's ratio are only used in the inherent strain FEM program **inhs2d.exe**, which are 200,000 MPa and 0.3, respectively, for steels.

List of Input Data

In the following line, the characters and digital data with a gray background are copies of the input data relating to the material properties. A brief description of the input data is given on the right side of the input line.

```
200000.0, 0.30.        ←Card6: Young's modulus and Poisson's ratio
```

6.3.2.3 Input Data for Boundary Conditions

- The flag of the stress and strain state is JPPD = −1.
- JPPD = −1 indicates the plane stress state.

Symmetric Boundary Conditions

- As the x-axis is the symmetric axis shown in Fig. 6.17, displacements of nodes in the y direction along the x-axis are set to zero.

Constraint Condition for Rigid Body Movement

- To constrain the rigid body movement in the x direction, displacement in the x direction at the central node is set to zero.
- The total number of constrained nodal displacements is NFIX = 302.
- The computation type is NCASE = 1.
- NCASE = 1 means to compute the residual stresses using known inherent strains.

List of Input Data

In the following lines, the characters and digital data with a gray background are copies of the input data relating to the boundary conditions. Brief descriptions of the input data are given on the right side of each input line.

−1	←Card7: JPPD
302	←Card8: NFIX
3151, 1, 0,	←Card9: KFIXP, KFIXD, FIXDIS
1, 2, 0,	←Card9: continue
22, 2, 0,	←Card9: continue
: : : : : : :	←Card9: skip print
6301, 2, 0,	←Card9: last input of constraint BC
1	←Card10: NCASE

6.3.2.4 Input Data of Inherent Strains

Inherent Strain Components

- Three components ε_x^*, ε_y^*, and γ_{xy}^* are considered for the plane stress problem.

Value of Inherent Strains

The value of the inherent strains in all elements is set to be equal to the residual plastic strains computed by the thermal elastic-plastic FEM program in Section 6.2, which was saved in the result file **pstrain.txt**. The format of the result file **pstrain.txt** is the same format as the input card11 and card12 of the input file **inhs2d.inp**.

List of Input Data

In the following lines, the characters and digital data with a gray background are copies of the input data relating to the known inherent strains in all elements. Brief descriptions of the input data are given on the right side of each input line.

$ELEMENT STRAIN-PX STRAIN-PY STRAIN-PXY	←Card11: comment line
1 0.305009E-02 -0.102563E-01 0.582277E-02	←Card12: Elem.id and $(\varepsilon_x^*, \varepsilon_y^*, \gamma_{xy}^*)$

```
2 0.135305E-02 -0.116922E-01 0.552450E-02      ←Card12: continue
:::::::::::::::::::::::::::::::::::::::::         ←Card12: skip print
5999 0.000000E+00 0.000000E+00 0.000000E+00    ←Card12: continue
6000 0.000000E+00 0.000000E+00 0.000000E+00    ←Card12: end
```

6.3.2.5 Save the Input Data

After the input data is prepared, save the input file **inhs2d.inp** in the computation folder as follows: **DATA13\inhs2d.inp**.

6.3.3 Steps to Execute the Program

This procedure can be performed using the following steps:

1. Create a computation folder called **DATA13**.
2. Copy the prepared input data from **DATA13\inhs2d.inp** to **DATA13**.
3. Copy the program to **DATA13**.
4. Double-click the program **inhs2d.exe** to start the computation.

6.3.4 Comparison of the Results of the Inherent Strain Method and the Thermal Elastic-Plastic Method

6.3.4.1 Distribution of Welding Shrinkage Deformation

Figures 6.18(a) and (b) show the distribution of longitudinal shrinkage ΔL along the transverse direction and the distribution of transverse shrinkage ΔS along the welding direction, respectively. In the figures, the solid line (**tepc**) and circles (**inhs**) indicate the results computed using the thermal elastic-plastic creep FEM program and reproduced by the inherent strain FEM program, respectively. The welding shrinkage ΔL and ΔS are reproduced by the inherent strain method with very good accuracy if the results computed by the thermal elastic-plastic method are considered as a standard for comparison.

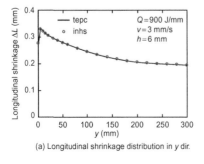

(a) Longitudinal shrinkage distribution in y dir.

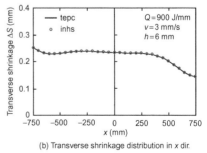

(b) Transverse shrinkage distribution in x dir.

FIG. 6.18 Reproduced longitudinal and transverse shrinkages by the inherent strain method.

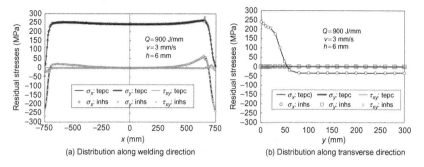

(a) Distribution along welding direction

(b) Distribution along transverse direction

FIG. 6.19 Reproduced residual stresses by the inherent strain method.

6.3.4.2 Distribution of Residual Stresses

Figures 6.19(a) and (b) show the distribution of residual stresses along the welding direction (x) and transverse direction (y), respectively. In the figures, tepc and inhs indicate the results computed by the thermal elastic-plastic creep FEM program and reproduced by the inherent strain FEM program, respectively. The results produced by the inherent strain method are the same as the results computed by the thermal elastic-plastic method.

6.4 NUMERICAL EXPERIMENT FOR RESIDUAL STRESS MEASUREMENT USING THE INHERENT STRAIN FEM PROGRAM

6.4.1 Purpose and Simulation Conditions

6.4.1.1 Purpose

- To learn the procedures for preparing the input data **inhs2d.inp** for the measurement of welding residual stresses.
- To learn how to use the inherent strain method and FEM program **inhs2d. exe** to estimate welding residual stresses.
- To confirm the accuracy of welding residual stresses and their distribution using the inherent strain method.

6.4.1.2 Object

- The measuring object is a butt-welded joint 1500 mm in length, 60 mm in width, and 6 mm in thickness, as shown in Fig. 6.20. The x direction is the welding line direction and the y direction is the transverse direction.

6.4.1.3 Conditions

Measured Strains

- Measured strains are assumed to be equal to the elastic strains (ε_x^e, ε_y^e) computed by the thermal elastic-plastic FEM. The measuring points are selected so that the total number of strains is more than the total number of unknown inherent strains.

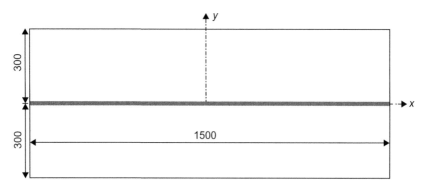

FIG. 6.20 Butt joint for the measurement of welding residual stresses.

Residual Stresses

- Residual stresses can be computed by the elastic FEM by applying the inherent strains as the initial strains to the object.

6.4.1.4 User Manual

The user manual is saved in **Manual\inhs2d_manual.pdf**.

6.4.2 Preparation of Input File

Based on the manual, **Manual\inhs2d_manual.pdf**, the following input data (card0–card17) must be prepared and saved in data file **inhs2d.inp**.

6.4.2.1 Input Data for Mesh Division

Mesh Division

Considering the symmetric geometry of a butt-welded joint, only half of the joint is necessary to create the mesh division, as shown in Fig. 6.21. In this example, the measured strains are assumed to be equal to the computed residual elastic strains. Therefore, the mesh division for stress measurement and thermal elastic-plastic analysis is the same. In the actual measurement, it

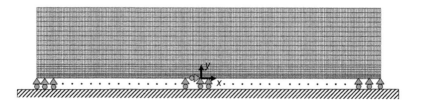

FIG. 6.21 Mesh division and boundary conditions.

is recommended to make the element center and element size fit the strain gauge center and strain gauge size.

List of Input Data

In the following lines, the characters and digital data with a gray background are copies of the input data relating to the mesh division. Brief descriptions of the input data are given on the right side of each input line.

$Inherent strain method model of butt welding	←Card0: comment
6321	←Card1: total node number
1-750.00000000000 0.0000000000000 0.0000000000000	←Card2: node id and cord
2-750.00000000000 5.0000000000000 0.0000000000000	←Card2: continue
3-750.00000000000 10.000000000000 0.0000000000000	←Card2: continue
∷∷∷∷∷∷∷∷∷∷∷∷∷∷	←Card2: skip print
6321 750.00000000000 300.00000000000 0.0000000000000	←Card2: end node input
6000	←Card3: total element no.
1 1 22 23 2 1	←Card4: eid, pid, and nodes
2 1 43 44 23 22	←Card4: continue
3 1 64 65 44 43	←Card4: continue
∷∷∷∷∷∷∷∷∷∷∷∷∷∷	←Card4: skip print
6000 2 6299 6320 6321 6300	←Card4: end elem. input
6.000	←Card5: thickness

6.4.2.2 Input Data for Material Properties

In the following line, the characters and digital data with a gray background are copies of the input data relating to the material properties. A brief description of the input data is given on the right side of the input line.

200000.0, 0.30,	←Card6: Young's modules, Poisson's ratio

6.4.2.3 Input Data for Boundary Conditions

- Flag of stress and strain state: JPPD = −1.
- JPPD = −1 indicates the plan stress state.

Symmetric Boundary Conditions

As shown in Fig. 6.21, the x-axis is the symmetric axis, and the y displacement of nodes in the x-axis is set to zero.

Constraint Condition for Rigid Body Movement

- To constrain the rigid body movement in the x direction, the x displacement at the central node is set to zero.
- The total number of constrained nodal displacements is NFIX = 302.
- The computation type is NCASE = 3.
  If NCASE = 3, the program **inhs2d.exe** will first compute the inherent strain from the measured strains and then continuously compute the residual stress using the inherent strains. If NCASE = 2, the program **inhs2d.exe** will only compute the inherent strain.

List of Input Data

In the following lines, the characters and digital data with a gray background are copies of the input data relating to the boundary conditions. Brief descriptions of the input data are given on the right side of each input line.

```
−1              ←Card7: JPPD
302             ←Card8: NFIX
3151, 1, 0,     ←Card9: KFIXP, KFIXD, FIXDIS
1, 2, 0,        ←Card9: continue
22, 2, 0,       ←Card9: continue
: : : : : :     ←Card9: print skipped
6301, 2, 0,     ←Card9: end NFIX
3               ←Card10: computation type NCASE
```

6.4.2.4 Input Data for Setting Unknown Inherent Strains

Inherent Strain Zone

The inherent strain zone is assumed to be the same as the plastic strain zone produced in welding. If the zone is more widely defined than the plastic strain zone, there is no problem for computation, but the total number of unknown inherent strains existing in the elements within the inherent strain zone will increase. In this example of a numerical experiment, the inherent strain zone is set based on the plastic strain zone computed by the thermal elastic-plastic analysis in Section 6.2. Figure 6.22 shows the inherent strain zone enclosed by a broken line. The half width is about 55 mm from the weld line. There

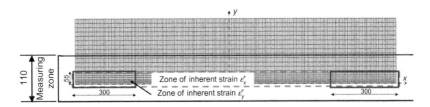

FIG. 6.22 Effective inherent strain zone and elastic strain measuring zone.

are 300 elements in the weld line direction and 8 elements in the transverse direction with $y < 55$ mm within the inherent strain zone. In total, there are 2400 elements in the inherent strain zone.

Inherent Strain Components

There are three inherent strain components $(\varepsilon_x^*, \varepsilon_y^*, \gamma_{xy}^*)$ for the plane stress problem. The inherent strain ε_x^* and its values in all the elements within the inherent strain zone are effective inherent strain for producing residual stresses as described in Chapter 1. The inherent strain ε_y^* is distributed linearly along the weld line, and it turns out that transverse shrinkage does not produce residual stresses. It becomes noneffective inherent strain. However, the inherent strain ε_y^* nonlinearly distributed along the weld line does produce both residual stresses and deformation. Therefore, the nonlinearly distributed inherent strain ε_y^* is the effective inherent strain. Many research results have shown that the nonlinearly distributed inherent strain ε_y^* is only located near the two ends of the butt-weld line and the shear inherent strain γ_{xy}^* is negligibly small.

Total Unknown Number of Inherent Strains

As shown in Fig. 6.22, there are 8 elements in the y direction for the half width of the inherent strain zone, being 55 mm and 300 elements in the weld line direction. Therefore, the total number of unknown inherent strain components ε_x^* is 2400. The effective inherent strain ε_y^* exists within only 300 mm from both ends of the weld line. The total element number with the existence of ε_y^* is 960. The total unknown number of effective inherent strains ε_x^* and ε_y^* is 3360, that is, **NINHA = 3360**.

Independent Inherent Strains

If the inherent strain in several elements is the same, the independent unknown number become less and then the required number of measuring elastic strains may be decreased. There is a flag **inhflg** for the independent inherent strains in the input data for the FEM program **inhs2d.exe**. In this example, all inherent strain components are independent, and the flag **inhflg** is set to zero in card12. Card14 and card15 are not necessary if **inhflg** is zero.

Card13 in the input data is the element number and the direction number for the component of inherent strains, and the direction number of component ε_x^* is **kdinha = 1**.

List of Input Data

In the following lines, the characters and digital data with a gray background are copies of the input data relating to the definition of unknown inherent strains. Brief descriptions of the input data are given on the right side of each input line.

$Inherent strain or measured strain	←Card11: comments
3360, 0,	←Card12: NINHA, inhflg
1, 1	←Card13: keinha, kdinha

2, 1	←Card13: continue
::::::::	←Card13: continue
2400, 1	←Card13: continue
1, 2	←Card13: continue
::::::::	←Card13: continue
2400, 2	←Card13: end card13

6.4.2.5 Input Data for Setting Measured Strains

Measuring zone and Number of Measured Strains

The total number of measured strains must be more than the total unknown number of inherent strains. In this numerical experiment, the measuring zone is assumed to be from $y = 0$ to $y = 110$ mm in the transverse direction. There are 3600 elements in the measuring zone. If two normal strains ε_x^m and ε_y^m are assumed to be measured in an element, the total number of measured strains can be 7200, that is, **NMEAS = 7200**.

Measuring Methods for Elastic Strains

The elastic strains can be measured by strain gauges attached to measuring positions or nondestructive measuring methods such as an X-ray method. In this numerical experiment, the elastic strains computed by the thermal elastic-plastic creep FEM program are assumed to be measured strains. If the measured stresses calculated from the measured strains are used as input data, there is a flag variable **nstrn** for stress input or strain input. **nstrn = 1** and **nstrn = 2** indicate the strain input and stress input, respectively. In this example, **nstrn = 1**.

List of Input Data

In the following lines, the characters and digital data with a gray background are copies of the input data relating to the measured elastic strains. Brief descriptions of the input data are given on the right side of each input line. For an example, the **ke**, **kd**, and **smeas** in the second line of the following input data are the element number, the direction number of the strain component, and the value of elastic strain, respectively.

7200, 1	←Card16: nmeas and mstrn
1, 1, 3.56E-04	←Card17: ke, kd, smeas
2, 1, 4.19E-04	←Card17: continue
::::::::	←Card17: print skip
3600, 2, 5.46E-05	←Card17: end

6.4.2.6 Save Input Data

After the input data is prepared, save the input file **inhs2d.inp** in the computation folder as follows: **DATA14\inhs2d.inp**.

6.4.3 Steps to Execute the FEM Program for Residual Stress and Deformation

This procedure can be performed using the following steps:
1. Create a computation folder called **DATA14**.
2. Copy the prepared input data from **DATA14\inhs2d.inp** to **DATA14**.
3. Copy the program to **DATA14**.
4. Double-click the program **inhs2d.exe** to start the computation.

6.4.4 Comparison of the Results of the Inherent Strain Method and the Thermal Elastic-Plastic Method

The inherent strains identified by inverse computation using measured strains are shown in Fig. 6.23. Figures 6.23(a) and (b) indicate the distributions of inherent strains ε_x^* and ε_y^* along the welding direction and on the middle transverse section, respectively. The plastic strains ε_x^p denoted by **epx:tepc** computed by the thermal elastic-plastic creep FEM program are also plotted in the same figures for comparison. The inherent strains identified by the measured strains are completely the same as the plastic strains. The inherent strains have a complicated distribution near the two ends of the weld line.

The residual stresses reproduced by these inherent strains are shown in Fig. 6.24. Figures 6.24(a) and (b) are the distributions of residual stresses along the welding line

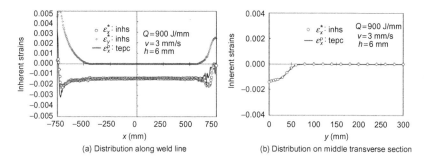

(a) Distribution along weld line   (b) Distribution on middle transverse section

FIG. 6.23 Distributions of inherent strains ε_x^* and ε_y^*.

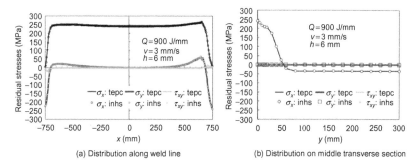

(a) Distribution along weld line   (b) Distribution on middle transverse section

FIG. 6.24 Distributions of residual stresses σ_x and σ_y.

direction and the middle transverse direction. tepc and inhs in the figure are the results using the thermal elastic-plastic program and the inherent strain program, respectively. The residual stresses reproduced by the inherent strain method are the same as those by the thermal elastic-plastic method including the residual stresses near the two ends of the weld line.

6.5 COMPUTATION STEPS FOR THE PREDICTION OF RESIDUAL STRESSES BY THE INHERENT STRAIN METHOD

6.5.1 Purpose and Simulation Conditions

6.5.1.1 Purpose

- There are two purposes of this section. One is to explain the relation between the inherent strain and welding conditions including material properties. Another is to explain the steps for predicting welding residual stresses by the inherent strain method.

6.5.1.2 Object

- The object is a butt-welded joint 1500 mm in length, 60 mm in width, and 6 mm in thickness, as shown in Fig. 6.25. The x direction is in the welding line direction and the y direction is in the transverse direction.

6.5.2 Preparation of Input File and Prediction Formula of Inherent Strain

Based on the manual, **Manual\inhs2d_manual.pdf**, the following input data (card0–card17) must be prepared and saved in data file **inhs2d.inp**. The inherent strains are prepared using the prediction formula.

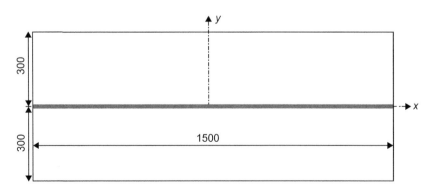

FIG. 6.25 Butt joint for prediction of welding residual stresses.

6.5.2.1 *Input Data for Mesh Division*

If the inherent strain is known, the computation of welding residual stresses is just to perform a simple elastic FEM analysis. Any of the uniform meshes or nonuniform meshes can be used. Here, the mesh shown schematically in Fig. 6.26 is adopted in this computation.

In the following lines, the characters and digital data with a gray background are copies of the input data relating to the mesh division. Brief descriptions of the input data are given on the right side of each input line.

```
$Inherent strain method model of butt welding    ←Card0: comment
6321                                              ←Card1: total node number
1-750.00000000000 0.0000000000000 0.0000000000000 ←Card2: node
                                                      id and cord.
2-750.00000000000 5.0000000000000 0.0000000000000 ←Card2:
                                                      continue
3-750.00000000000 10.000000000000 0.0000000000000 ←Card2:
                                                      continue
::::::::::::::::::::::::::                         ←Card2: skip print
6321 750.00000000000 300.00000000000 0.0000000000000 ←Card2: end
                                                      card2
6000                                              ←Card3: total elem. number
1 1 22 23 2 1                                     ←Card4: elem. id and node id
2 1 43 44 23 22                                   ←Card4: continue
3 1 64 65 44 43                                   ←Card4: continue
::::::::::::::::::::::::::                         ←Card4: skip print
6000 2 6299 6320 6321 6300                        ←Card4: end card4
6.000                                             ←Card5: thickness
```

6.5.2.2 *Input Data for Material Properties*

The material properties used in the inherent strain method are only Young's modulus and Poisson's ratio. The values of Young's modulus and Poisson's ratio are 2×10^5 MPa and 0.3, respectively.

In the following line, the characters and digital data with a gray background are copies of the input data relating to the material properties.

```
200000.0, 0.30,          ←Card6: Young's modulus and Poisson's ratio
```

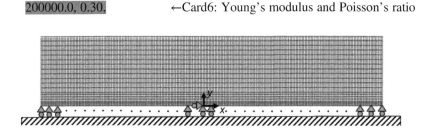

FIG. 6.26   Mesh division and boundary conditions.

6.5.2.3 Input Data for the Boundary Conditions

- The flag JPPD for the two-dimensional plane stress state is -1 (**JPPD = -1**).
- The y displacements of the nodes on the x-axis are constrained, as shown in Fig. 6.26.
- The x displacement at one node shown in the Fig. 6.26 is also constrained to prevent the rigid movement.
- The total number of constrained nodal displacements is NFIX = 302.
- The computation type is set to 1 (**NCASE = 1**) for the computation of the residual stress using the predicted strains.

In the following lines, the characters and digital data with a gray background are copies of the input data relating to the boundary conditions. Brief descriptions of the input data are given on the right side of each input line.

−1	←Card7: JPPD
302	←Card8: NFIX
3151, 1, 0,	←Card9: KFIXP, KFIXD, FIXDIS
1, 2, 0,	←Card9: continue
22, 2, 0,	←Card9: continue
: : : : : :	←Card9: skip print
6301, 2, 0,	←Card9: end card
1	←Card10: computation type flag NCASE

6.5.2.4 Prediction Formula of Inherent Strain

Inherent Strain Component

Generally, in the plane stress state of a butt weld, three inherent strain components ε_x^*, ε_y^*, and γ_{xy}^* must be considered for the prediction of residual stresses. If the weld line is long enough, only the longitudinal residual stress is produced except in the vicinities of the ends of the weld line. This implies that the inherent strain ε_x^* is a dominant component in the most part of the joint and the inherent strains ε_y^* and γ_{xy}^* are small and can be ignored.

Distribution Function of Inherent Strain

The distribution of the inherent strain ε_x^* along the welding direction and in the transverse section is schematically shown in Fig. 6.27. The same value of inherent strain ε_x^* in the welding direction can be used to predict the welding residual stresses. In the transverse direction, the pattern of distribution can be approximately expressed in the form of the function given by

$$\varepsilon_x^*(y) = \begin{cases} \varepsilon_{xW}^* & (0 \le y \le y_H) \\ \varepsilon_{xW}^* \cdot \dfrac{b-y}{b-y_H} & (y_H < y \le b) \end{cases} \tag{6.5.1}$$

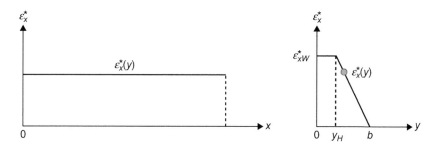

FIG. 6.27 Distribution of inherent strain ε_x^*.

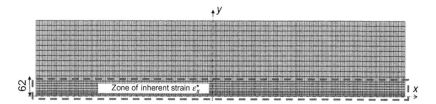

FIG. 6.28 Mesh division and inherent strain existing zone.

The y_H in Fig. 6.27 and Eq. (6.5.1) is the half width of the mechanical melting zone, which is approximately equal to the heat-affected zone (HAZ) for steel. The b is the half width of the existing inherent strain zone as shown in Fig. 6.28. In the zone $0 \leq y \leq y_H$, the inherent strain has the same value ε_{xW}^*. In the zone $y_H \leq y \leq b$, the inherent strain changes linearly. In the outside of the inherent existing zone $y \geq b$, the inherent strain is zero.

Parameters of Inherent Strain Distribution Function

There are three parameters y_H, b, and ε_{xW}^* in the inherent strain distribution function given by Eq. (6.5.1). They can be predicted by Eqs. (6.5.2), (6.5.3), and (6.5.4), respectively [1]:

$$y_H = \frac{Q}{\sqrt{2\pi}ec\rho h(T_m - T_0)} \tag{6.5.2}$$

$$b = \frac{\alpha Q}{\sqrt{2\pi}ec\rho h\varepsilon_{YB}}\left(1 - \frac{0.27\alpha \cdot T_{av}}{\varepsilon_{YB}}\right) \tag{6.5.3}$$

$$\varepsilon_{xW}^* = -\varepsilon_{YW}\left(1 + \frac{0.27\alpha \cdot T_{av}}{\varepsilon_{YB}}\right) \tag{6.5.4}$$

TABLE 6.3 Variables and Their Values Used in the Inherent Strain Prediction Formula

Variables	Meanings	Values
$c[J/(Kg°C)]$	Specific heat	6.594×10^2
$\rho[Kg/mm^3]$	Mass density	7.72×10^{-6}
$\alpha[1/°C]$	Thermal expansion coefficient	1×10^{-5}
ε_{YW}	Yield strain of welded metal	0.0012
ε_{YB}	Yield strain of base metal	0.001
$Q[J/mm]$	Effective heat input per weld line	900
$T_m[°C]$	Mechanical molten temperature	775
$T_0[°C]$	Initial or preheating temperature	20
$B[mm]$	Width of plate to be welded	300
$h[mm]$	Thickness	6

The T_{av} in Eqs. (6.5.2) and (6.5.3) is called the average temperature increase in the butt-welded joint given by Eq. (6.5.5), which is used to adjust the inherent strain distribution with consideration of the effect of difference in the size of the welded joint.

$$T_{av} = \frac{Q}{2Bh \cdot c\rho} \tag{6.5.5}$$

The meaning of the variables in the above equations (6.5.2)–(6.5.5) is described in Table 6.3. Their values for mild steels and welding conditions used in this chapter are shown on the right side of Table 6.3. Because the prediction equations ignore the change of the material properties with temperature, the recommended material properties are the average values from the root temperature to the molten temperature.

Generally, the yield stress of welded metal is higher than the base metal. Therefore, the yield strain ε_{YW} of the welded metal in Table 6.3 has a higher value than the yield strain ε_{YB} of the base metal.

Calculation of Inherent Strain Parameters

By substituting the values given in Table 6.3 into the equations (6.5.2)–(6.5.5), the average temperature increase T_{av} and the inherent strain parameters $y_H, b, \varepsilon_{xW}^*$ can be easily calculated as shown in Table 6.4.

TABLE 6.4 Inherent Strain Parameters and Their Values

Inherent Strain Parameters	Values
T_{av}: Average temperature increase	49.1°C
y_H: Half width of HAZ	9.44 mm
b: Half width of inherent strain zone	61.9 mm
ε_{xW}^*: Inherent strain value in welded zone	−0.001359

6.5.2.5 Input Data of Inherent Strain

The inherent strains will be applied to the elements for FEM computation. Therefore, the elements where inherent strain exists must be determined first. In calculating the half width ($b = 61.8$ mm) of the inherent strain zone, it can be easily shown that only 8 elements in the transverse direction from the weld line have inherent strain. There are 300 elements in the weld line direction. The total number of elements having the inherent strain is 2400.

By substituting the y coordinate of the center of the elements into Eq. (6.5.1), the digital value of inherent strain in the elements as input data can be calculated as follows:

- Inherent strain for element id from 1 to 300:

$$y = 2.5 \text{ mm}, \varepsilon_x^* = -0.001359.$$

- Inherent strain for element id from 301 to 600:

$$y = 7.5 \text{ mm}, \varepsilon_x^* = -0.001359.$$

- Inherent strain for element id from 601 to 900:

$$y = 12.5 \text{ mm}, \varepsilon_x^* = -0.001280.$$

- Inherent strain for element id from 901 to 120:

$$y = 17.5 \text{ mm}, \varepsilon_x^* = -0.001150.$$

- Inherent strain for element id from 1201 to 1500:

$$y = 25 \text{ mm}, \varepsilon_x^* = -0.000956.$$

- Inherent strain for element id from 1501 to 1800:

$$y = 35 \text{ mm}, \varepsilon_x^* = -0.000696.$$

- Inherent strain for element id from 1801 to 2100:

$$y = 45 \text{ mm}, \varepsilon_x^* = -0.000437.$$

- Inherent strain for element id from 2101 to 2400:

$$y = 55\,\text{mm}, \varepsilon_x^* = -0.000178.$$

- Inherent strain for element id from 2401 to 6000:

$$y > 55\,\text{mm}, \varepsilon_x^* = 0.$$

In the following lines, the characters and digital data with a gray background are copies of the input data relating to the predicted inherent strains in all elements. Brief descriptions of the input data are given on the right side of each input line.

$ Data of Inherent strain or measured strain	←Card11: comments
1, −0.001359, 0, 0	←Card12: element, $\varepsilon_x^*, \varepsilon_y^*, \gamma_{xy}^*$
2, −0.001359, 0, 0	←Card12: continue
: : : : : : : :	←Card12: skip print
5999, 0, 0, 0	←Card12: continue
6000, 0, 0, 0	←Card12: end card12

6.5.2.6 Save Input Data

After the input data is prepared, the input file **inhs2d.inp** should be saved to the computation folder as follows: **DATA15\inhs2d.inp**.

6.5.3 Steps to Execute inhs2d.exe for Residual Stress Computation

This procedure can be performed using the following steps:

1. Create a computation folder called **DATA15**.
2. Copy the prepared input data from **DATA15\inhs2d.inp** to **DATA15**.
3. Copy the program to **DATA15**.
4. Double-click the program **inhs2d.exe** to start the computation.

6.5.4 Comparison of the Results by the Inherent Strain Method and the Thermal Elastic-Plastic Method

The residual stresses predicted by the inherent strain method and the thermal elastic-plastic method are shown in Fig. 6.29. Figures 6.29(a) and (b) show the residual stress distributions along the weld line and on the middle transverse section, respectively. Near the middle transverse section, the residual stresses sx:inhs and sy:inhs predicted by the inherent strain ε_x^* are the same as the residual stresses sx:tepc and sy:tepc computed by the thermal elastic-plastic method. Near the two ends of the weld line, the predicted residual stresses have a similar distribution, but there are some differences in magnitude compared to the results using the thermal elastic-plastic method.

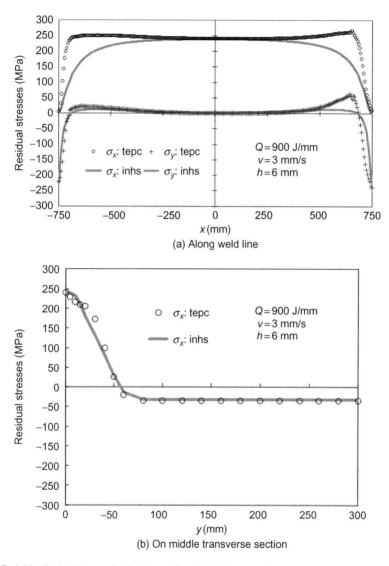

FIG. 6.29 Residual stress distribution predicted by inherent strain.

REFERENCE

[1] Ueda Y, Yuan MG. Prediction of residual stresses in butt welded plates using inherent strains, Journal of Engineering Materials and Technology. Trans ASME 1993;115:417–23.

Chapter 7

Strategic Simulation Analyses for Manufacturing Problems Related to Welding

In the preceding chapters, the basic theories of welding mechanics and elastic-plastic analysis by the finite element method (FEM) and the inherent strain method were presented. With the help of these methods, we can find solutions for manufacturing problems of welded joints and construction problems from a mechanical perspective.

There are two different approaches to dealing with manufacturing problems:

1. When encountering manufacturing problems, investigate the causes of the problem and establish countermeasures.
2. Prior to adopting a new manufacturing procedure, investigate it from various aspects such as by computational simulation, experiments, etc., to propose the best procedures and methods.

From the viewpoint of welding mechanics, the key to solving such problems is to minimize welding stress/strain and deformation. Generally speaking, when trying to minimize welding deformation, we have to apply a strong constraint that causes a larger magnitude of internal stress/strain. In contrast, less constraint causes more welding deformation but less internal stresses/strains. Accordingly, the fundamental goal should be how to control these contradictory factors.

When we simulate behavior by thermal elastic-plastic analysis, we take the following steps. First, we try to analyze the welding phenomenon under the condition as accurate as possible to the actual one. Next, we find points where large stresses/strains are induced. To develop preventive measures against the problems with large stress/strain, we perform simulation analyses, changing the welding conditions and manufacturing procedures to find the best measures. This simulation procedure is also applied when developing new manufacturing methods.

Since inherent stains are plastic strains that are produced during elastic-plastic behavior under a thermal cycle induced by welding, inherent strain is the result of complicated nonlinear behavior. As we have learned, the inherent strain method takes advantage of the simplicity of linear elastic analysis to predict welding residual stress and deformation, and further to measure residual stresses in two and three dimensions.

To develop some countermeasures against manufacturing problems and propose new manufacturing procedures, actual efforts are exerted on some of the examples presented in this chapter. In Appendix A, actual residual stresses, measured by the experiments and obtained by thermal elastic-plastic analysis, are illustrated, edited to supplement the examples in this chapter.

7.1 COLD CRACKING AT THE FIRST PASS OF A BUTT-WELDED JOINT UNDER MECHANICAL RESTRAINT

Cold cracking at the first pass of a butt-welded joint is a typical manufacturing problem. The main cause is local stress/strain produced during the cooling stage of the welding procedure, being subjected to the influence of diffusible hydrogen. To clarify the relationship between the cold cracking and the restrained stress/strain, a RRC (rigidly restrained cracking) test is sometimes used. With this test the relation between the external restraint of the specimen and the local stress/strain is also clarified. Accordingly, if we can estimate the degree of external restraint of the specimen, we can predict the local stress and propose some measure to prevent the cold cracking.

Here, we perform the thermal elastic-plastic analysis on the RRC test specimens in the entire course of welding to find where and how large local stresses are induced.

The RRC specimen of mild steel and its idealization for finite element analysis are represented in Figs. 7.1 and 7.2 [1–3]. The thermal elastic-plastic analysis is carried out for the first pass during the welding. The computed local stresses and strains in the x direction σ_x and ε_x are shown in Fig. 7.3 with focus to points A, B, C, and D, where large magnitudes are anticipated in relation to the cracking.

When welding is complete, the welded metal is rapidly cooled down by conduction to the base metal and it starts to shrink. The shrinkage of the weld metal

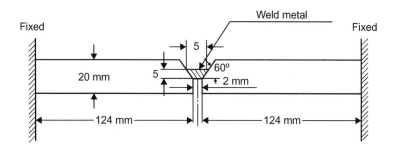

FIG. 7.1   RRC test specimen.

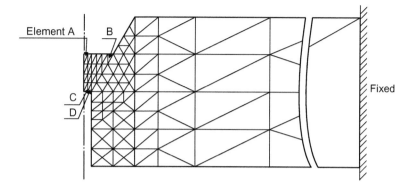

FIG. 7.2 Division by finite elements.

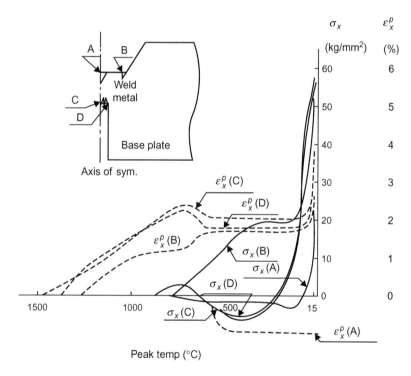

FIG. 7.3 Histories of stress and plastic strain in the RRC specimen.

is resisted by the base plate to induce tensile stress. The most stress is produced on the bottom side of the weld metal, points C and D, since the length of the bottom side is shorter than the upper side and results in a stronger constraint. At the second stage of cooling, the upper side is gradually cooled down and the shrinkage strain starts to increase at points A and B. At the same time, this causes a small amount of bending due to the eccentricity of the location of the welded metal, but this is also resisted by the base metal and results in a small amount of bending. At the final stage of cooling, a large amount of heat energy stored in the joint is emitted outside and the joint shrinks. The shrinkage at the final stage is resisted by the wall to produce large tensile stress, σ_x and ε_x, in the joint. The resistance of the wall is called the external constraint. The center line of the welded metal is located out of the center line of the base metal, and the gap between these center lines causes the bending, which induces additional tensile stress on the upper sides and compressive stress on the bottom sides of the welded metal. As previously discussed, the external constraint has two roles: axial and rotational constraints. The main role is the axial constraint; the degree of constraint becomes lower and the constrained stress becomes lower when the length of the joint becomes longer.

Distribution of residual stresses is denoted by the solid lines in Fig. 7.4. The integration of the axial residual stresses in the cross-section equals the restrained force at the wall. Even in the case of a specimen being free from the external constraint, residual stresses are produced, which are shown by the one-dot-chain lines in Fig. 7.4. Very large stress is detected in the welded metal, but this is compressive and should not initiate cracking.

As is generally accepted for conventional welded joints, a higher tensile residual stress/strain is produced proportionally to a higher intensity of restraint.

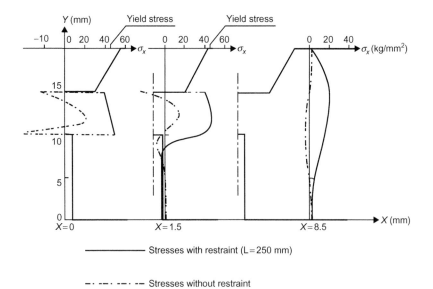

FIG. 7.4 Longitudinal stresses (σ_x) in the RRC specimen.

So, one of the important measures to prevent cracking is to reduce the intensity of the restraint of the joints.

7.2 COLD CRACKING OF SLIT WELD

In the process of manufacturing welded products, occasionally we have to apply welding to a slit. As the slit displays strong restraint against thermal expansion and shrinkage of the welded portion, it may induce cold cracking. To cope with this, the y-slit weld cracking test specimen, as shown in Fig. 7.5 [4], is often used to find the material quality, intensity of restraint, preheating temperature, etc., as to prevent this kind of cold cracking.

In slit welds of this type, cold cracks often initiate where the large residual stress/strain is induced. The residual stress/strain is produced by the thermal behavior of the specimen at the heating and cooling stages. At the heating stage by welding, the base plate along the slit is heated and expands, and the expansion suppresses the weld metal in the slit. At the cooling stage, the shrinkage of the base plate along the slit induces tension on the weld metal, which also starts to shrink by cooling. A combination of these two mechanical behaviors is the cause of residual stresses. The location of the largest magnitude of residual stress/strain changes depending on the ratio of these two behaviors, which is

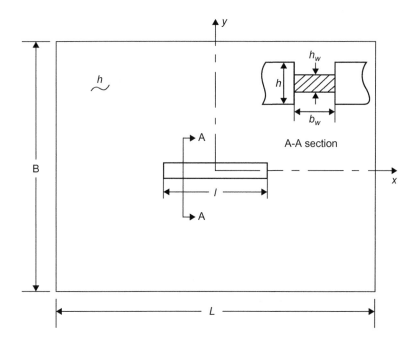

FIG. 7.5 Specimen with a slit.

influenced by the size ratio of the specimen, the heat input of welding, etc. In the following, thermal elastic-plastic analysis is performed to clarify the basic mechanical characteristics of the slit weld.

Usually, the degree of restraint intensity of a slit is defined as the elastic relation between the applied force and the resulting displacement along the slit, or vice versa. The main reason the restraint intensity is utilized is because the residual stress is considered to be approximately proportional to it. However, in the case where slit welding is applied to the y-slit specimen, the resulting distribution of the residual stress/strain is different from that predicted by the restraint intensity. As a matter of fact, the residual stress in the standard specimen reaches the yield stress along the entire slit length and the plastic strain is greater in the middle than near the ends of the slit. This suggests that cold cracking may initiate in the middle rather than near the ends. This result contradicts the prediction by the intensity of restraint.

When we observe the mechanical behavior of a slit weld during welding based on the thermal elastic-plastic analysis, we realize that the source of welding stress is composed of two components: shrinkage of the weld metal and displacement of the base plate along the slit [5]. The latter component is dominant so that the former may be neglected. Taking this fact into account, an elastic-plastic analysis for a slit weld was developed [6]. With this method, a series of analysis is carried out to clarify the general characteristics of the residual stress/strain relation of slit welds.

The base specimen is the y-slit weld specimen (breadth: $B = 150$ mm; length: $L = 200$ mm; length of the slit: $\ell = 80$ mm). The analysis is performed on similar slit specimens of which the geometry is similar to the base one, but the sizes are different. The welding conditions are as follows:

- plate thickness: $h = 20$ mm
- throat thickness: $h_w = 5$ mm
- gap of the slit: $b_w = 4$ mm
- yield stress of weld metal: $\sigma_Y = 50$ kgf/mm^2
- heat input of welding: $Q = 17{,}000$ J/cm (instantaneous heat input)
- rigidity recovery temperature: $T_m = 700°C$
- initial temperature: $T_0 = 15°C$
- specific heat: $c = 0.188$ cal/g·°C
- density: $\rho = 7.66$ g/cm^3
- efficiency of heat input: $\varepsilon = 0.75$
- linear expansion coefficient: $\alpha = 1.2 \times 10^{-5}$ (1/°C)

The residual stress/strain distributions illustrated in Table 7.1 are classified into three groups according to the magnitude:

1. Elastic distribution (stress state (1) in Table 7.1). In a longer slit length (smaller heat input against the slit length), the residual stress of the weld metal is in the elastic range except in the vicinities of the ends.

TABLE 7.1 Characteristics of Restrained Stress and Strain Induced in Welded Metal of a Specimen with a Slit

Stress State	Heat Input Against Slit Length		Restrained Stress	Restrained pl Strain	Total Restrained el and pl Strains
(1)	Small		σ_Y	—	
(2)	Medium		σ_Y		
(3)	Large	(a) S	σ_Y		
		(b) M			
		(c) L			

2. Elastic-plastic distribution (stress state (2) in Table 7.1). As the slit length becomes shorter, the plastic region is extended from the ends to the middle and plastic strain is induced in a short range from the ends.
3. Plastic distribution over the entire slit length (stress state (3) in Table 7.1).

In the case where the slit length becomes much shorter, the residual stress of the weld metal is in the plastic range along the entire slit length (stress state (3)). Further, this stress state may be divided into three groups **a**, **b**, and **c** in view of the distribution of resulting plastic strain ε_w^p, as follows:

a. The pattern of distribution of ε_w^r is ⊔ and the severe range of intensity of restraint is expanded toward the middle.
b. The pattern of distribution of ε_w^r is approximately elliptic and the restraint is high in the middle portion of the slit.
c. The pattern of distribution of ε_w^r is ⊓ and the restraint is the most severe in the middle.

According to the conventional intensity of restraint, the condition becomes more severe toward the slit ends. In contrast to this, the magnitude of ε_w^p becomes greater in the middle in the cases of (b) and (c) in stress state (3). This indicates that the restraint condition is more severe in the middle than near the ends. The y-slit cracking test specimen belongs to this category.

Judging from these results, the conventional intensity of restraint is not necessarily the best mechanical index against weld cracking, and a new index has been proposed taking into consideration the plastic restrained strain [7].

With this information, we can anticipate where we have the cold cracking of the first weld, in the middle or at the ends of the slit, referring to the size of specimen and the welding condition.

7.3 ANALYSIS OF WELDING RESIDUAL STRESS OF FILLET WELDS FOR PREVENTION OF FATIGUE CRACKS

In the construction of steel structures, fillet welding is indispensable for connecting structural components, etc. The fillet weld is required to have sufficient fatigue strength as well as static strength. For initiation of fatigue cracks, welding residual stresses is influential and so is the geometrical configuration of the fillet weld. Special attention should be paid to the fillet weld by boxing. To cope with this concern, a series of thermal elastic-plastic analysis is performed under various welding conditions. Here, part of the results [8–10] is presented. From the analysis, it is also found that two-dimensional analysis is sufficiently accurate to estimate residual stresses in the middle portion of a long welded joint.

7.3.1 Residual Stresses by Three-Dimensional Analysis

In the referred paper [8], the thermal elastic-plastic analysis is conducted on continuous fillet straight welds and fillet welds by boxing.

Figure 7.6 shows the specimens of the analysis, which are composed of flanges 16 mm thick and webs 12 mm thick, and jointed by a single-pass weld.

The material of the joint is of high tensile strength steel, of which the yield stress is 400 MPa at room temperature. Figure 7.7 displays the finite element mesh division by three-dimensional solid elements.

The results of the analysis are shown in Figs. 7.8(a)–(d). In each figure, the residual stresses are on Line 1 along the longitudinal weld line, Line 2 at the weld by boxing, Line 3 in the transverse section, and Line 4 in the longitudinal section, respectively. In Fig. 7.8(d), the residual stresses (test) obtained by the experiment are shown to compare with the analyzed one (analysis) and to demonstrate the accuracy of the analysis. Figure 7.8(b) indicates that large tensile stress is induced at the weld by boxing at A in the figure, especially the component σ_x, which is at the yield stress. This is one important factor for reducing the fatigue strength of a joint.

7.3.2 Comparison of Residual Stresses by Two-Dimensional and Three-Dimensional Analyses

In the previous section, the three-dimensional analysis was conducted using three-dimensional solid elements. As this computation requires a large computation time, it is not always convenient to carry out this series of calculations under various welding conditions [9,10]. In contrast, residual stress is almost always uniform

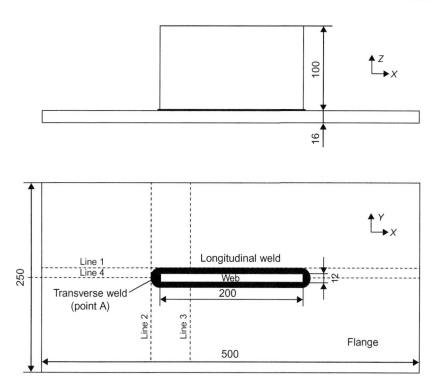

FIG. 7.6 Configuration of specimen with fillet welds.

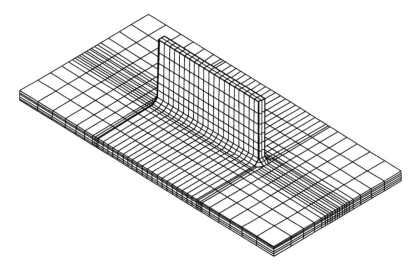

FIG. 7.7 Division of specimen by three-dimensional solid finite elements.

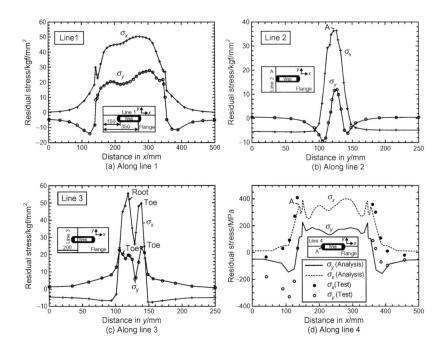

FIG. 7.8 Welding residual stresses in fillet welds analyzed by three-dimensional solid finite elements.

in the middle part of a long welded joint except near both ends. This suggests that two-dimensional analysis is a useful means only if the residual stresses at the transverse section in the middle part are of concern. To this point, two-dimensional and three-dimensional analyses are conducted on a T-joint connected by a fillet weld. The mesh divisions by three-dimensional solid elements and two-dimensional plate elements are illustrated in Figs. 7.9 and 7.10, respectively.

The two-dimensional analysis is conducted satisfying the generalized plane strain condition, which is exhibited by

$$\varepsilon_x(y, z) = a_1 + a_2 y + a_3 z. \tag{7.3.1}$$

In Eq. (7.3.1), a_1, a_2, and a_3 are coefficients determined from the two conditions: the resultant of axial stresses on the cross-section should vanish, $Fx = 0$, and the resultants of moments should be zero, $My = Mz = 0$.

In Fig. 7.11(a) the longitudinal distributions of σ_x and σ_y are obtained by the three-dimensional thermal elastic-plastic analysis. The distributions are almost uniform along the length except in the vicinities of both ends. In Fig. 7.11(b) the transverse distributions of σ_x, σ_y, and σ_z in the middle of the length are displayed, which are calculated by the three-dimensional FEM and two-dimensional FEM with the generalized plane strain condition. The results of these calculations show similar distributions with reasonable accuracy. This fact indicates that two-dimensional analysis can be utilized effectively in some instances.

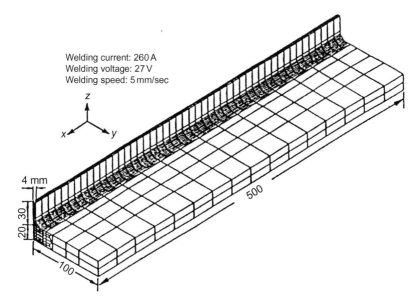

FIG. 7.9   Division of fillet welds by three-dimensional solid elements.

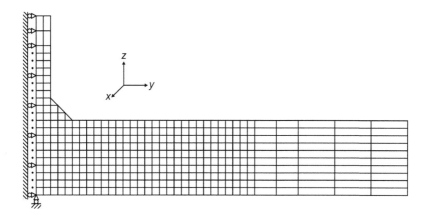

FIG. 7.10   Division of transverse section of fillet welds by two-dimensional plate elements.

7.3.3   Comparison of Residual Stresses in Single-Pass and Multipass Welds

Figure 7.12 represents a cross-section of a multipass-welded joint, which is prepared by simultaneous welding at both sides [9,10]. To this joint, two-dimensional thermal elastic-plastic analysis is conducted under the generalized plane strain condition in two cases of single-pass and three-pass welds. In Fig. 7.13, the

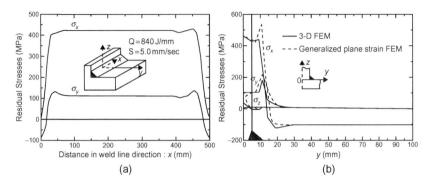

FIG. 7.11 Comparison of residual stresses calculated by three-dimensional and two-dimensional analyses.

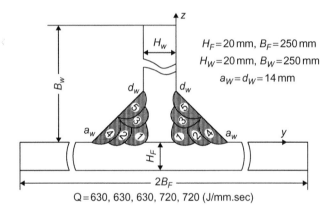

FIG. 7.12 Multipass fillet welds and welding condition.

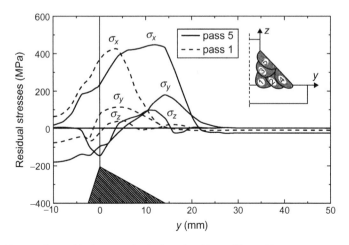

FIG. 7.13 Welding residual stresses in single and multipass fillet welds.

resulting residual stress distributions induced on the surface of the flange are shown. The patterns of the distributions are similar and the maximum stress is larger in the multipass weld and its location moves toward the toe.

7.4 MULTIPASS-WELDED CORNER JOINTS AND WELD CRACKING

Multipass-welded corner joints are often used in bridges, industrial machines, and other constructions. At these joints, in addition to root cracking, open-type lamellar tearing (i.e., tearing that reaches to the surface) is occasionally observed at the position outside the heat-affected zone (HAZ). Occurrence of these two types of weld cracking is largely dependent on the degree of restraint against flexural welding deformation of corner joints. To develop measures against these types of weld cracks, an investigation was carried out by theoretical analysis and experiment. The result [11–14] is introduced in the following section.

7.4.1 Experiment and Result

In this investigation, the CJC (corner joint weld cracking) test apparatus [12] was developed to clarify the effect of bending restraint intensity on lamellar tearing and root cracking in multipass-welded corner joints, as illustrated in Fig. 7.14. The CJC test specimen consists of two plates, each 100 mm wide. The plates are held down by rollers so that the angular distortion of each plate at the inner end of the restraint length l is completely restrained, but they are allowed to move axially. The intensity of bending restraint K_B is evaluated by the equation shown in Fig. 7.14.

A steel plate of 40 mm thickness is used. The material is 50 kgf/mm^2 (490 N/mm^2) class, high tensile–strength steel. Test welds were made using manual arc metal (MMA) welding under the condition shown in Table 7.2. The throat depth was 20 mm, which was prepared by five layers with nine passes. The thermal cycle was measured at the center of the weld line on the top surface of the vertical plate by thermal couple, as shown in Fig. 7.14. For measurement of the welding residual stress, a thin plate (9 mm thick) was cut out from the middle portion of the specimen vertically with respect to the weld line. The measurement was performed by a sectioning method, putting strain gauges on the thin plate. Then, the observed residual stresses are in a plane stress state.

7.4.2 Residual Stresses by Thermal Elastic-Plastic Analysis

From the previous experimental result, it was seen that transverse welding stress is the most influential factor of the occurrence of weld cracks. Reflecting on the

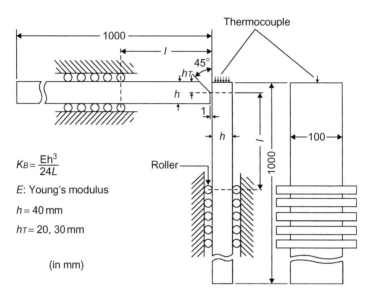

FIG. 7.14 CJC test apparatus.

TABLE 7.2 Welding Conditions

Welding Process	Preheating Temp.	Interpass Temp.	Heat Input	Build-up Process
Covered arc welding	50°C	50°C	22,000 J/cm	$h_T = 20\,mm$

experimental condition, the analysis is conducted on the transverse section with unit weld length (throat depth 20 mm) in the plane stress state [11]. Heat conduction analysis is conducted, assuming each weld pass is an instantaneous heat source with 80% efficiency of heat input (details are shown in Table 7.2). The finite element idealization is illustrated in Fig. 7.15. The temperature-dependent physical properties are from [11].

Introducing the results of the heat conduction analysis, thermal elastic-plastic analysis is carried out, taking into consideration the temperature-dependent

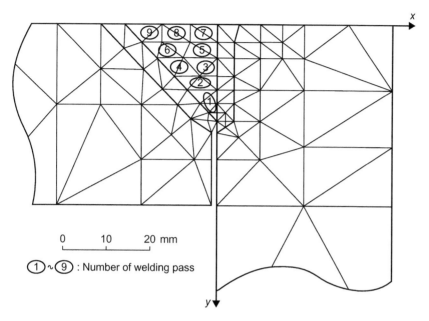

FIG. 7.15 Division by finite elements for analysis.

mechanical properties of the material. In the analysis, two external restraint conditions are adopted: the large intensity of restraint $K_B = 1000 \times 10^3$ kgf/m·m·
rad ($l = 56$ mm), denoted by $K_B - 1000$, and nonrestraint condition ($K_B - 0$).
In multipass welding, the effect of the subsequent passes on the weld metal is considered in the analysis [11].

7.4.3 Effects of Welding Residual Stress and Geometry of Edge Preparation on Initiation of Welding Cracks

As for temperature distributions of the joints during welding, the results of heat
conduction analysis correlate well with those of the experiments. This suggests
that theoretical analysis is conducted under the proper welding condition and its
results should be reliable. Using the results of the analysis, the thermal elastic-
plastic analysis is conducted.

Next, the welding stresses are presented. Figure 7.16 indicates the transverse
stress σ_x to the weld line at several interpass temperatures (the 9th pass at room
temperature) in the case of $K_B - 1000$. In Fig. 7.16(a), large tensile stresses on
the upper end surface of the vertical plate are observed apart from the toe

between the 6th to 9th interpass temperatures. This implies the stresses may cause lamellar tearing. Figure 7.16(b) shows the welding stresses along the HAZ, which are compressive at the 2nd pass. This indicates the stresses do not induce root cracking.

The residual stresses shown in Fig. 7.17 indicate some difference between the experiment and theoretical analysis. As the experimental specimen is sized by gas cutting, this causes residual stress around the top surface. Then, the final residual stress is produced by gas cutting and welding. The residual stress caused by gas cutting is measured and presented in Fig. 7.18.

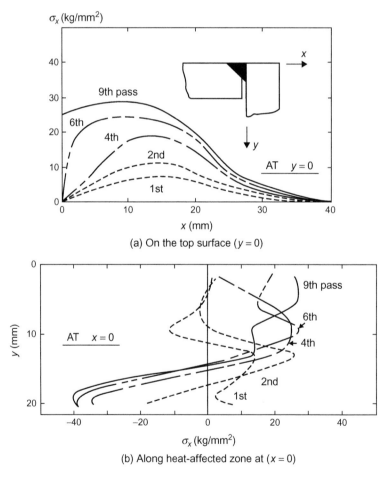

FIG. 7.16 Transverse welding transient and residual stresses; σ_x (analysis). ($K_B = 1000 \times 10^3$ kgf · mm/mm · rad.)

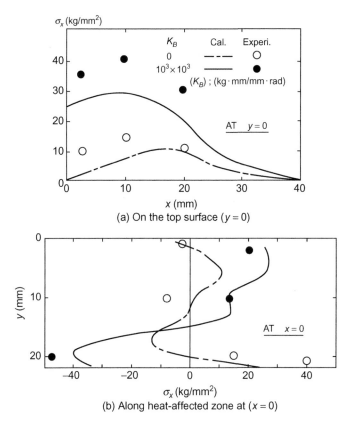

FIG. 7.17 Transverse welding residual stresses; σ_x (analysis and experiment). ($K_B = 1000 \times 10^3$ kgf · mm/mm · rad.)

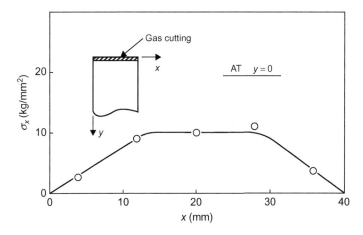

FIG. 7.18 Residual stress induced by gas cutting (at $y = 0$).

Type	G	M	P	W	C
Shape of groove	G (Gas cut) h hT $\leftarrow h \rightarrow$	M (Mechanical cut)	Δ_0 G	M	M $h/2$ h

$h = 40\,\text{mm}, hT = 20, 30\,\text{mm}, \text{Angle of vee}: 45°, \Delta_0 = 10\,\text{mm}$

FIG. 7.19   Shapes of groves used in CJC test.

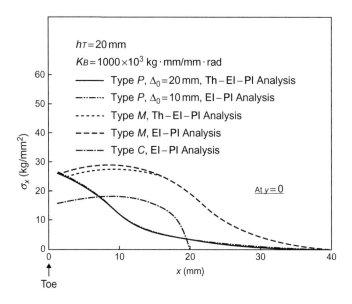

FIG. 7.20   Transverse welding residual stress distributions analyzed theoretically.

It is recognized that the residual stress is about 10 kgf/mm² in the middle portion of the plate thickness ($x = 10$–$30\,\text{mm}$) [11]. Taking this into consideration, the experimental and analyzed residual stresses correlate well with each other. The geometry of edge preparation also influences the weld cracks. The measurement and analysis on the welding residual stresses are conducted on the corner joints of the different types of edge preparation shown in Fig. 7.19. The results are represented in Figs. 7.20 and 7.21 [13,14], and provide very useful information for prevention of the cracking at corner joints.

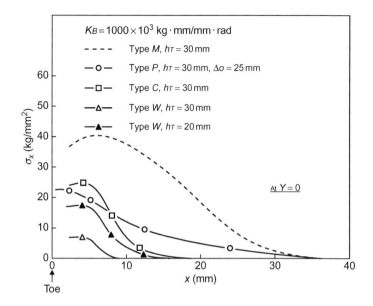

FIG. 7.21 Transverse welding residual stress distributions measured experimentally.

7.5 ANALYSIS OF TRANSIENT AND RESIDUAL STRESSES OF MULTIPASS WELDING OF THICK PLATES IN RELATION TO COLD CRACKS, UNDER-BEAD CRACKS, ETC.

Pressure vessels of high-quality thick plates are the main structural components of chemical plants, nuclear reactors, etc. These thick plates are usually joined by multipass welding. By this welding, root cracks and/or under-bead cracks may occur by the transient and residual stresses [15]. To prevent these cracks, stress-relief annealing is sometimes required, which may cause other types of cracks (stress-relief cracks). One of the main influential factors is the magnitude of these stresses, which closely relates to the degree of restraint against welding deformation. To clarify the production process of the welding stresses, a series of thermal elastic-plastic analysis is performed on simplified specimens with a narrow gap [16].

7.5.1 Specimens and Conditions for Theoretical Analysis

In Fig. 7.22(a), Model 1 of the simplified specimen is shown, which is 50 mm in thickness and 95 mm in width, with a narrow gap with a groove width of 10 mm. Each layer is formed by one-pass welding, and 20 passes are laid on this Model 1. The welding condition is the same for all models: heat input is 30,000 J/cm and its efficiency is 95%. The preheat temperature and interpass temperature are set to room temperature, 15°C. As the model is symmetric with respect to the middle cross-plane (y-z plane), the analysis is performed on half of the model.

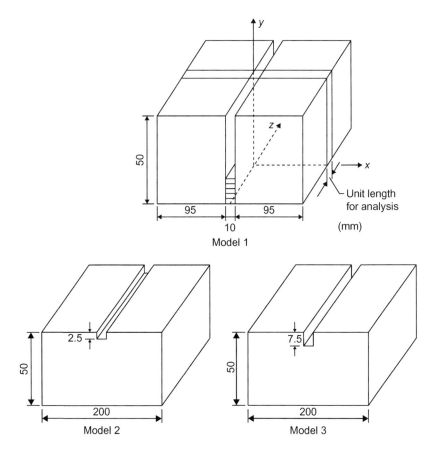

FIG. 7.22 Specimens with a narrow gap for analysis.

In the analysis, the region where the maximum temperature reaches above 700°C is regarded as the HAZ.

The analysis is conducted under two different restraint conditions: conditions A and B, as shown in Figs. 7.23(a) and (b), respectively. Condition A imposes no external restraint on the model during welding. Condition B restricts the rotation of the model about the x-axis and z-axis (longitudinal deformation and angular distortion), but the model is free in the x direction. The degree of restraint of the actual models is considered to be between these two extreme conditions. Although the model is sufficiently long, every pass is assumed to be formed instantaneously by welded metal. Accordingly, it might be rational to regard that the y-x plane vertical to the weld line is allowed to move and rotate, remaining as a plane.

Under these conditions, three-dimensional thermal elastic-plastic analysis is conducted. The physical and mechanical properties of the metal (SM50) used in the analysis are shown in Figs. 2 and 3 in [16].

The analysis is carried out on Model 1 as accurately as possible. So, the mesh division shown in Fig. 7.24 is fine and one pass is composed of four meshes.

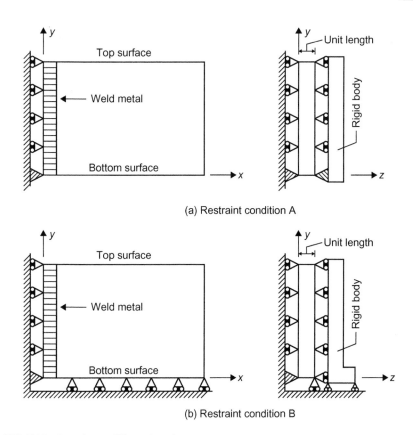

(a) Restraint condition A

(b) Restraint condition B

FIG. 7.23 Restraint conditions of specimens.

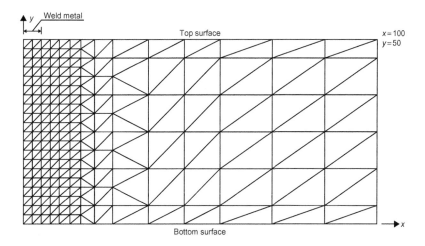

FIG. 7.24 Mesh division for analysis.

For the 20 passes equivalent to 20 layers, the analysis is carried out for each pass successively.

7.5.2 Characteristics of Welding Residual Stress Distributions and Production Process

The results of the foregoing analysis are presented in Figs. 7.25(a)–(d). In Figs. 7.25(a) and (b), the transverse residual stress σ_x and longitudinal stress σ_z in the middle plane, $x = 0$, and those on the top surface ($y = 50$ mm) in Figs. 7.25(c) and (d) are illustrated. According to the results, the residual stress in the thickness direction σ_y and the shear stress τ_{xy} are small compared to the maximum values of those stresses σ_x and σ_z.

The production mechanism of the residual stresses is discussed in the following, starting with those produced under restraint condition B.

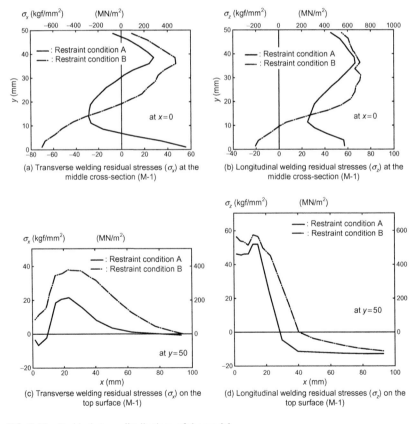

(a) Transverse welding residual stresses (σ_x) at the middle cross-section (M-1)

(b) Longitudinal welding residual stresses (σ_z) at the middle cross-section (M-1)

(c) Transverse welding residual stresses (σ_x) on the top surface (M-1)

(d) Longitudinal welding residual stresses (σ_z) on the top surface (M-1)

FIG. 7.25 Residual stress distributions of the model.

7.5.2.1 Condition B

Under this condition, the longitudinal and transverse bending deformations are restrained. Then, the residual stress is produced by restraint against the shrinkage of the welded metal and HAZ of each pass. Although the shrinkage is resisted by the accumulated welded zone below the newly laid pass, it adds compressive stress to the restraining portion in return. With repetition, the entire distribution of residual stress is formed. As a result, the compressive σ_x and σ_z are induced near the bottom surface.

7.5.2.2 Condition A

Under this condition, the specimen is completely free from the external restraint. Then, longitudinal and transverse bending deformations of the specimen occur by the multipass welding, which produce additional stresses to those induced under condition B. Consequently, σ_x and σ_z are large in tension near the bottom surface. It is well recognized that this tensile residual stress may cause root cracks. In actual welded joints, however, the restraint condition is considered between the two extreme conditions A and B. Then, the large tensile residual stresses may not be produced at the bottom surface.

In contrast to this, large tensile stresses, σ_x and σ_z, appear near the top surface under both conditions A and B, as noted in the figures. Their maximum values are observed a few layers below the finishing bead and should be near the exterior boundary of the HAZ. The maximum value in each restraint condition is the largest over the entire distribution.

As the model of this analysis is idealized by a large number of meshes, the result should be sufficiently accurate, but the analysis requires a large amount of computer processing unit power. To save computation time, several trial calculations are shown in [15]. The result suggests using large mesh division except in the important areas. For example, if attention is focused near the top surface, the models illustrated in Figs. 7.22(b) and (c) may be used for analyses and experiments.

7.6 IMPROVEMENT OF RESIDUAL STRESSES OF A CIRCUMFERENTIAL JOINT OF A PIPE BY HEAT-SINK WELDING

Many stainless steel pipes such as SUS 304 are used in various plants such as nuclear reactors because of their excellent performance in severe circumstances. Nevertheless, intergranular stress corrosion cracking may occur on the inner surface of the welded zone if the pipe is under (1) large tensile residual stress, (2) high sensitivity of material, and (3) environment of high oxygen content together [17]. To prevent this stress corrosion cracking, one of the mechanical measures is to create the welding residual stresses on the inner surface to compression. One such method is the heat-sink welding by which the inner surface of the welded zone is compulsorily cooled by water-spraying during welding [17,18].

Applying this heat-sink welding and conventional welding, the experiment and FEM analysis were conducted on the circumferential multipass-butt welding of SUS 304 4B pipes, as indicated in Fig. 7.26, to verify the validity of the method [17].

By conventional welding, the specimens are naturally cooled down after welding. In the heat-sink welding, the first three passes are done by conventional welding and for the following 4th to 6th passes, the inner surface of the welded zone is sprayed with cooling water (room temperature) at a flow rate of 22 L/min. Measured and analyzed residual stresses produced by the conventional welding and heat-sink method are presented in Figs. 7.27 and 7.28, respectively. Both results are well correlated to each other, and the effect of the heat-sink method is noted to produce compressive stress on the inner surface of the pipes. Similar experiments and analyses were performed on pipes of 2B, 4B, and 24B under different welding conditions. The results [18] are illustrated in Section A.9.1 of Appendix A.

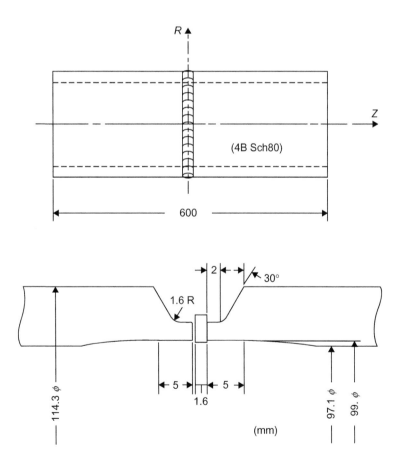

FIG. 7.26 Pipe specimen used for experiment (4B pipe) and details of groove preparation for joint.

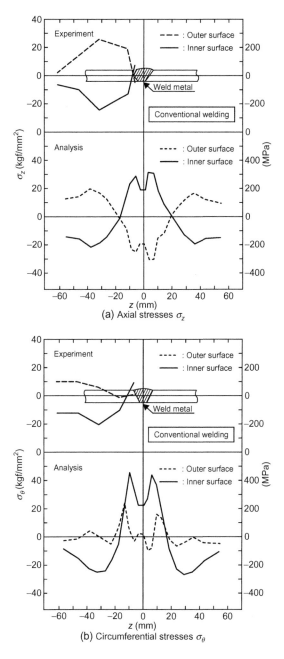

FIG. 7.27 Residual stress distributions on inner and outer surfaces of pipe by experiment and analysis (in case of conventional welding).

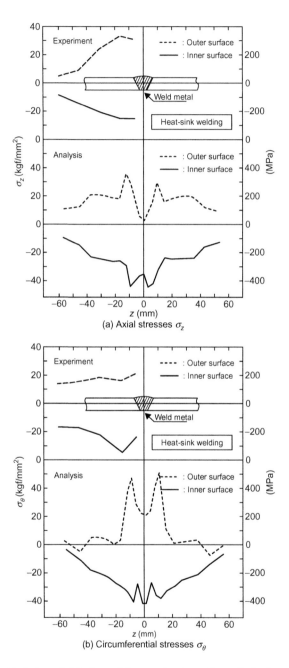

FIG. 7.28 Residual stress distributions of inner and outer surfaces of pipe by experiment and analysis (in case of heat-sink welding).

7.7 PREDICTION OF DEFORMATION PRODUCED BY LINE HEATING

Ships are built by assembling curved steel plates. To form these curved plates, line heating is employed. In line heating, the same mechanism that produces welding distortion is used. When the steel plate is heated along a line in the welding, local deformation, such as transverse shrinkage, longitudinal shrinkage, and angular distortion, are produced. If these local deformations are given to a flat plate in a controlled manner, a curved plate with an arbitrary shape can be formed. This is the basic idea of plate forming by line heating. The same line heating can be used to straighten welding deformations. The difference between welding and line heating is the temperature and the heat source. In welding, the steel plate is heated above the melting temperature, which is around 1450°C, while in line heating, the temperature is kept below 800°C. An arc is used in welding and gas flame or induction heating is used in line heating.

Figure 7.29 shows the typical types of curved plates used in ship building [19]. The pillow shape, the saddle shape, and the twisted shape are three basic shapes, and a combination of these is also found. The curved surface can be divided into developable and undevelopable surfaces. The former can be produced by only bending the plate, and both bending and stretching or shrinkage are necessary to produce the latter. Cylinders and corns are developable surfaces while the sphere is undevelopable.

According to this definition, all the curved surfaces shown in Fig. 7.29 are undevelopable surfaces. The strain that is necessary to form the curved surfaces can be computed by FEM elastic analysis to deform the curved plate to a flat plate [20–22]. Figure 7.29 shows the distribution of the bending strain and

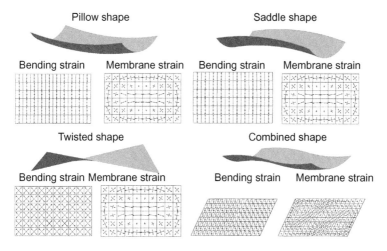

FIG. 7.29 Typical types of curved surfaces and bending and membrane strains necessary to form these.

the membrane strain necessary to form the four typical curved surfaces. Comparing the bending and the membrane strains, the distribution of the bending strain has simple and relatively uniform distribution, while that of the membrane strain is very complex. This is the reason plate forming requires experienced workers. The knowledge of experienced workers may be appreciated from Fig. 7.29. It is seen from the figure that both the pillow and the saddle shape have a similar bending strain distribution. The pillow shape is produced by giving shrinkage along the edge of the plate, while the saddle shape is produced by giving shrinkage to the center part of the plate.

If the relation between the strain produced by a single heating line and the heating condition is accumulated as a database, it becomes possible to replace the distributed strain distribution by a group of discrete heating lines. Based on this idea, a plan for the plate forming by line heating can be automatically generated.

7.8 SIMULATION OF RESISTANCE SPOT WELDING PROCESS

Welding simulation, which traces the evolution of deformation and stress during welding, can be used not only to predict welding distortion and residual stress, it can also be used to simulate weldability. In this section, its application to resistance spot welding is presented as an example. As shown in Fig. 7.30, two or more than two plates to be joined are squeezed by a pair of electrodes

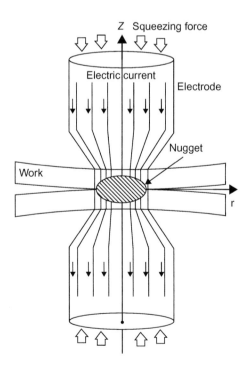

FIG. 7.30 Schema of resistance spot welding.

and a large current is given between these electrodes in the resistance spot welding. The plates are melted by Joule heat. The shaded area in the figure is the melted part, called the nugget. When the works are cooled to the room temperature, the joining process is completed.

The evolution of the distortion and the stress in spot welding can be analyzed as a thermal elastic-plastic problem if the transient temperature is known [23–30]. The difference from the arc welding and the thermal cutting is that the heat is mostly given as Joule heat in case of the resistance spot welding. In Joule heat, the rate of heat generation per unit volume is proportional to the square of the current density. Thus, the distribution of the current density must be known to compute the temperature field. Since the current density is strongly dependent on the path through which the current flows, the contact area between the electrode and the work or those between the works must be traced accurately. To simulate such a problem by the FEM, the change of electric, thermal, and mechanical fields in a small time increment are computed cyclically, as shown in Fig. 7.31. Though the details of the theory are not discussed here, the mechanical analysis and the electric field analysis belong to the type of problem consisting of a spring only, as shown in Table 4.2 in Chapter 4. Similarly, the thermal analysis belongs to the type of problem consisting of a spring and a damper. All these problems can be solved by the FEM.

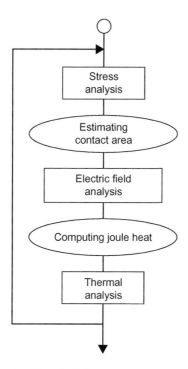

FIG. 7.31   Flowchart of spot welding simulation.

As one of the examples, the influence of the electrode shape on the nugget formation in aluminum alloy plates studied by using the FEM is presented. The electrodes considered are the R-type and the dorm-type electrodes, as shown in Fig. 7.32. In the R-type, the tip has a spherical surface with a radius of 100 mm. The contact area increases with the increase of the squeezing force. The tip of the dorm-type electrode within 6 mm from the axis is a sphere with a radius of 40 mm and the corner of the tip surface has R with an 8 mm radius. Therefore, the contact area increases with the squeezing force until the spherical tip reaches the full contact, and the contact area does not increase beyond this limit.

In this example, the thickness of the plate is 1 mm. Figures 7.33 and 7.34 show the nugget growth when the squeezing force is 2500 N, and the current changes from 15,000 A to 25,000 A (60 HZ AC current). Comparing the two

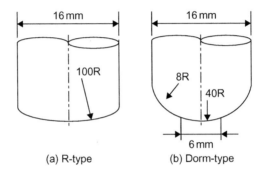

FIG. 7.32   Shape of R-type and Dorm-type electrode.

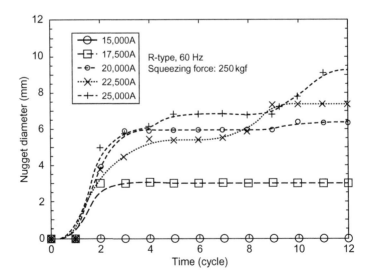

FIG. 7.33   Time history of nugget growth (R-type electrode).

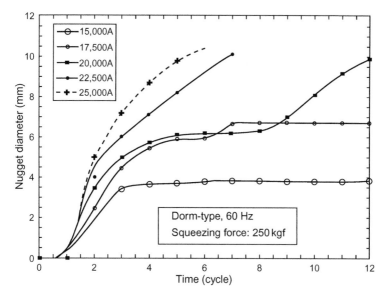

FIG. 7.34 Time history of nugget growth (Dorm-type electrode).

electrodes, the nugget stops to grow at the 4th cycle when the electrode is the R-type. In the dorm-type electrode, the nugget is kept growing even after four cycles when the current is large. Such difference is attributed to the difference of the electrode shape. An R-type electrode promotes the nugget formation in the early stage of welding. With the increase of the contact area due to the softening of the work between the electrodes, the current density decreases and the heat input and nugget growth are controlled. In the dorm-type electrode, the contact area stops increasing after reaching the full contact of the electrode, and as a consequence the nugget keeps growing.

As demonstrated, the influence of various factors, such as the electrode shape, the control of the squeezing force, and the current in time on the nugget formation, can be studied using FEM simulation. It is also possible to get directions from FEM simulation to search the range of welding conditions to achieve sound spot welding joints.

7.9 PREDICTION OF WELDING DISTORTION PRODUCED IN LARGE PLATE STRUCTURES

Most welded structures are large in scale. Thermal elastic-plastic analysis cannot be a practical method for predicting welding distortion of such structures because of the unrealistically large computation time. On the other hand, the inherent strain method in which the welding distortion is estimated by elastic analysis using the inherent strain as the initial strain is advantageous in computation time. The irreversible strain produced by the welding is the inherent strain and its integration is the inherent deformation. When the welding line is long enough, inherent deformations, such as the transverse and the longitudinal shrinkages, distribute uniformly along the welding line except at the starting and finishing ends. Since their magnitude is mostly determined by the heat input and the plate thickness when the material is the same, the inherent deformation can be saved as a database.

In this section, prediction of the distortion of a curved stiffened plate structure under welding assembly is presented as an example. As shown in Fig. 7.35, a curved plate stiffened by two transverse and three longitudinal stiffeners is considered. The radius of curvatures of the curved plate in two principle directions x', y' are 7200 mm and 3600 mm, respectively. The angle θ between the longitudinal stiffener and the principal direction is 20°. Though inherent strain or inherent deformation is one of the major causes of distortion, the gap and the misalignment accumulated during assembly and their correction process are also important causes [31–33]. In this example, the influence of both the local deformations and the gaps are considered. As shown in Fig. 7.36(a), a 5 mm initial gap between the skin plate and the longitudinal stiffeners at the center is assumed. In the first stage, the gap is closed by the fitting and then the welding is performed. To clarify the influence of various types of mismatch, Case 1 (without gap), Case 2 (with initial out-of-plane deformation of the stiffener), Case 3 (with initial twisting deformation of the stiffener), Case 4 (with initial deflection in both out-of-plane and twisting directions), Case 5 (with the center gap caused by the initial in-plane deformation of the stiffener), and Case 6 (with the end gap caused by the initial in-plane deformation of the stiffener) are given. Figure 7.36(b) shows the distortion produced by the gap correction in Case 5. Noting that three corners are fixed in vertical direction, the lowest corner in the figure deformed upward. The same corner deforms downward by the welding, as shown in Fig. 7.36(c). The influences of various types of initial deformation on the distortion are summarized in Fig. 7.37. In Case 1

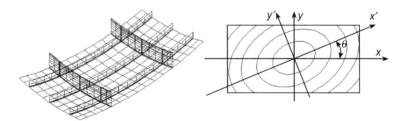

Bottom place: $L = 3000$ mm, $B = 1500$ mm, $t = 9$ mm
Longitudinal stiffener: $L = 3000$ mm, $H = 125$ mm, $t = 7$ mm
Transverse stiffener: $L = 1500$ mm, $H = 300$ mm, $t = 9$ mm

FIG. 7.35 Unsymmetric curved structure model.

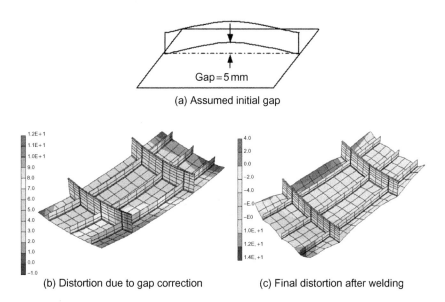

(a) Assumed initial gap

Gap = 5 mm

(b) Distortion due to gap correction   (c) Final distortion after welding

FIG. 7.36 Influence of gap on welding distortion of unsymmetric curved structure.

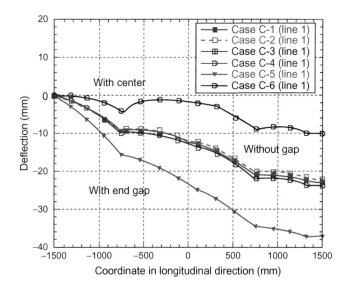

FIG. 7.37 Influence of gap between parts on geometrical accuracy in welding assembly.

without initial deformation, the free corner deforms downward by 23 mm due to the welding. The influence of the initial deformation of the stiffener in the out-of-plane and twisting direction is small, and the distortion by welding is dominant. Compared to this, that of the initial in-plane deformation such as in Case 5 and Case 6 is large. The gap correction process produced additional deformation and its magnitude is about 15 mm, which is larger than the size of the given gap. It is also interesting to note that the direction of the distortion changes whether the gap is located at the center or the end.

As shown by this simple example, simulation of the distortion produced by the welding assembly is effective for identifying which factor is influential and to control the dimensional accuracy of welded structures.

To validate the inherent deformation method, an experiment was conducted using the model shown in Fig. 7.38. The model is a plate structure stiffened by two transverse stiffeners and three longitudinal stiffeners. To use the inherent deformation method, inherent deformations, such as the transverse shrinkage, the longitudinal shrinkage and the angular distortion at each welding joint, must be known beforehand. These inherent deformations were estimated both by experiments and thermal elastic-plastic FEM using the welding joint models shown in Fig. 7.38. The joint models L-S, T-S and L-T are the joints between the longitudinal stiffener and the skin plate, between the transverse stiffener and the skin plate and between the longitudinal and the transverse stiffeners, respectively. The computed deflection of the structure without an initial gap is shown in Fig. 7.39. Further, the deflection along Line 1 in the transverse direction and Line 2 in the longitudinal direction is shown in Fig. 7.40. In this figure, the computed and the measured results are compared. For both the computation

FIG. 7.38 Plate structure for measurement of welding distortion and joint models to estimate inherent deformations.

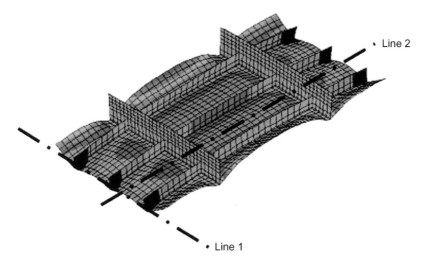

FIG. 7.39   Predicted welding distortion (without initial gap).

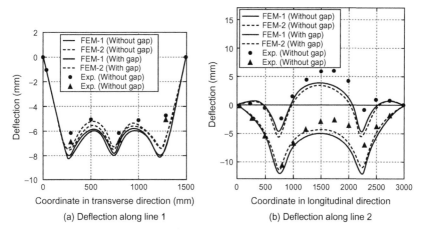

FIG. 7.40   Comparison of deflections predicted by the inherent deformation method and measurement.

and the measurement, cases without an initial gap and those with a 10 mm edge gap are shown. The lines in the figure show the computed results and the plotted points show the measured values. Solid triangles and solid circles represent the cases with and without the initial gap. The influence of gap correction is directly observed in the deflection along Line 2, but it is not clearly observed along Line 1. Comparing the computed and the measured results, the difference is about 2 mm. This demonstrates the effectiveness of the inherent deformation method.

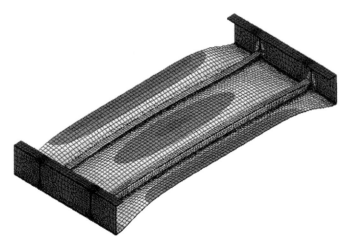

FIG. 7.41   Prediction of welding distortion produced in large structure.

As discussed, the inherent deformation method is an advantageous method for predicting the welding distortion of large welded structures. The thermal elastic-plastic FEM [33] can also be employed as a practical tool to predict the welding distortion of a fairly large structure, as shown in Fig. 7.41. The length of the structure in the figure is 3500 mm, and it is modeled using 46,000 8-node hexagonal elements. The computation took 34 hours using a 64-bit PC. Considering the remarkable progress of the capability of computers, the thermal elastic-plastic FEM will become a practical tool to predict welding distortion of large structures in the near future.

REFERENCES

[1] Ueda Y, Kusachi Y. Theoretical analysis of local stresses and strains in RRC test specimens at crack initiation. 1972. IIW Doc.X-662-72.

[2] Yamakawa T. Application of FEM to nonlinear structural problems. Dissertation. Osaka Univ., Japan; 1971.

[3] Ueda Y, Fukuda K, Nakacho K. Analysis of welding stress by FEM and mechanism of production of residual stresses. Jl Japan Welding Society 1976;45(1):29–35.

[4] Watanabe S, Satoh K. Welding mechanics and its application. Asakura Shoten, Tokyo; (in Japanese); 1965.

[5] Ueda Y, Fukuda K, Low J. Mechanism of production of residual stress due to slit weld. Trans JWRI 1974;3(2):159–66.

[6] Ueda Y, Fukuda K, Kim Y. Analytical calculation method of restraint stresses and strains due to slit weld in rectangular plates. Trans JWRI 1982;11(1):105–13.

[7] Ueda Y, Fukuda K, Kim Y, Koki R. Characteristics of restraint stress-strain of slit weld in a finite rectangular plate and the significance of restraint intensities as a dynamical measure. Trans JWRI 1982;11(2):105–13.

[8] Wu A, Ma NX, Murakawa H, Ueda Y. Effect of welding procedures on residual stresses of T-Joints. Trans JWRI 1996;25(1):81–89.

[9] Wu A, Ma NX, Murakawa H, Maeda H. FEM analysis of three-dimensional welding residual stresses and angular distortion in T-type fillet welds. Trans JWRI 1995;24(2):115–22.

[10] Ma NX, Ueda Y, Murakawa H, Yuan MG. FEM analysis on welding residual stresses in T-type fillet welds. In: Mathematical Modeling of Weld Phenomena. 3rd Ed. H. Cerjak, The Institute of Materials; 1997 (pp. 590–605).

[11] Ueda Y, Fukuda K, Nishimura I, Iiyama H, Chiba N. Dynamical characteristics of weld cracking in multipass welded corner joint. Trans JWS 1977;8(2):1–5.

[12] Ueda Y, Nishimura I, Iiyama H, Chiba N. Effect of intensity of bending restraint on lamellar tearing and root cracking in corner joint. Trans JWS 1977;8(2):122–29.

[13] Ueda Y, Nishimura I, Iiyama H, Chiba N, Fukuda K. Prevention of lamellar tearing in multipass welded corner joint. Trans JWS 1978;9(2):128–33.

[14] Ueda Y, Fukuda K, Nishimura I, Iiyama H, Chiba N. Cracking in welded corner joints. Jl of Metal Construction 1984;16(1):30–34.

[15] Ueda Y, Takahashi E, Fukuda K, Nakacho K. Transient and residual stresses in multipass welded. Trans JWRI 1974;3(1):59–67.

[16] Ueda Y, Nakacho K. Simplifying methods for analysis of transient and residual stresses and deformations due to multipass welding. Trans JWRI 1982;11(1):95–103.

[17] Ueda Y, Nakacho K, Shimizu T. Improvement of residual stresses of circumferential joint of pipe by heat-sink welding. Trans ASME Jl Press Vessel Technol 1986;108:14–22.

[18] Ueda Y, Nakacho K, Shimizu T, Ohkubo K. Residual stresses and their mechanisms of production at circumferential weld by heat-sink welding. J JWS 1983;52(2):90–97 (in Japanese).

[19] Tango Y, Nagahara S, Kobayashi J, Ishiyama M, Nagashima T. Automated line heating for plate forming by IHI-ALPHA system and its application to construction of actual vessels (System Outline and Application Record to Date). J Soc Naval Architects Japan 2003;2003:193.

[20] Ueda Y, Murakawa H, Rashwan AM, Okumoto Y, Kamichika R. Development of computer-aided process planning system for plate bending by line heating (Report 1: Relation between Final Form of Plate and Inherent Strain). Trans SNAME J Ship Production Feb. 1994;10(1):59–67.

[21] Ueda Y, Murakawa H, Rashwan AM, Okumoto Y, Kamichika R. Development of computer-aided process planning system for plate bending by line heating (Report 2: Practice for Plate Bending in Shipyard Viewed from Aspect of Inherent Strain). Trans SNAME J Ship Production. Nov. 1994;10(4):239–47.

[22] Ueda Y, Murakawa H, Rashwan AM, Neki I, Kamichika R, Ishiyama M, Ogawa J. Development of computer-aided process planning system for plate bending by line heating (Report 3: Relation between Heating Condition and Deformation). Trans SNAME J Ship Production Nov. 1994;10(4):248–56.

[23] Nied HA. The finite element modeling of the resistance spot welding process. Welding Research. Supplement Apr. 1984;123–32.

[24] Tsai CL, Dai WL, Dickinson DW, Paritan JW. Analysis and development of a real-time control methodology in resistance spot welding. Welding Research Supplement Dec. 1991;339–51.

[25] Tsai CL, Jammal OA, Papritan JC, Dickinson DW. Modeling of resistance spot weld nugget growth. Welding Research Supplement Feb. 1992;47–54.

[26] Wu KC. Resistance spot welding of high contact-resistance surfaces for weldbonding. Welding Research Supplement Dec. 1975;436–43.

[27] Zacharia T, Aramayo GA. Modeling of thermal stresses in welds. In: Proc. of International Conference on Modeling and Control of Joining Processes. Orlando, Dec. 1993;533–40.

[28] Syed M, Sheppard SD. Computer simulation of resistance spot welding as a coupled electrical-thermal-mechanical problem. In: Proc. of International Conference on Modeling and Control of Joining Processes. Orlando, Dec. 1993;422–29.

[29] Murakawa H, Kimura F, Ueda Y. Weldability analysis of spot welding on aluminum using FEM. Trans JWRI 1995;24(1):101–11.

[30] Deng D, Murakawa H, Liang W. Numerical simulation of welding distortion in large structures. Computer Methods in Applied Mechanics and Engineering 2007;196(45–48):15, 4613–27.

[31] Deng D, Murakawa H, Liang W. Prediction of welding distortion in a curved plate structure by means of elastic FEM. Journal of Materials Processing Technology 2008;203(1–3):18, 252–66.

[32] Deng D, Murakawa H, Shibahara M. Investigations on welding distortion in an asymmetrical curved block by means of numerical simulation technology and experimental method. Computational Materials Science 2010;48(1):187–94.

[33] Nishikawa H, Serizawa H, Murakawa H. Actual application of FEM to analysis of large scale mechanical problem in welding. Science and Technology of Welding and Joining 2007; 12(2):147–52.

Residual Stress Distributions in Typical Welded Joints

LIST OF RESIDUAL STRESS DISTRIBUTIONS IN TYPICAL WELDED JOINTS

Section	Base Plate or Welded Joints	Relevant Section	References
A.1	Residual stresses in base metals		
A.1.1	Residual stress in TMCP steel		[1,2]
A.1.2	Residual stress in TMCP steel induced by bead weld		[1,2]
A.1.3	Explosive clad steel		[3]
A.1.4	Cylindrical thick plate by cold bending		[4]
A.2	Residual stresses in welded joints of plates; 2-dimensional		
A.2.1	Butt-welded joints; classification of patterns of residual stress distributions		[5,6]
A.2.2	Long butt-welded joint, prediction equation		[7]
A.2.3	Built-up members of T shape and I shape		[7]
A.2.4	Built-up member of T shape, experiment		[8]
A.2.5	Residual stress and inherent displacement induced by slit welds	7.2	[9]
A.3	Multipass butt welds of thick plates; 3-dimensional		
A.3.1	Multipass butt welds of thick plates, classification	7.5	[10]

(Continued)

Section	Base Plate or Welded Joints	Relevant Section	References
A.3.2	Multipass butt welds of thick plates, experiment		[11]
A.4	Electron beam welds, thick plate		[12]
A.5	First bead of butt joint; RRC test specimen	7.1	[13,14]
A.6	Multipass-welded corner joint	7.4	[15]
A.7	Fillet welds; 3-dimensional		
A.7.1	Single fillet welds		[16]
A.7.2	Fillet welds at the joint of web and flange	7.3	[17,18]
A.8	Repair weld of thick plate; 3-dimensional		[19]
A.9	Circumferential welded joint of pipes		
A.9.1	Circumferential welded joint of pipes; heat sink welding	7.6	[20]
A.9.2	Penetrating pipe joints in nuclear reactor		[21–23]

A.1 RESIDUAL STRESSES IN BASE METALS

A.1.1 Residual Stress in Thermo-Mechanical Control Process (TMCP) Steel

Doc. No. 1.1	Base Metal of TMCP [1,2]		
Production	TMCP	Medium Thick	Material: High Strength Steel
Experiment	Measurement by L_x specimen and L_y specimen using stress release method.		
Components	σ_x = in rolling direction, σ_y = in transverse direction.		
Characteristics	Tensile residual stress in the middle of thickness, compressive one on upper and lower surfaces of plate.		

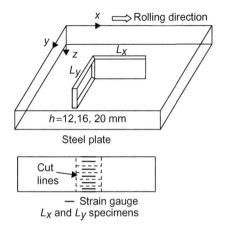

FIG. A.1 TMCP plate for measurement.

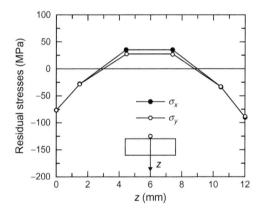

FIG. A.2 Residual stresses of TMCP (12 mm thick).

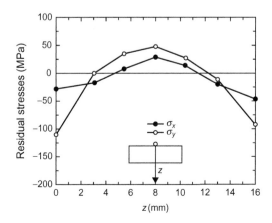

FIG. A.3 Residual stresses of TMCP (16 mm thick).

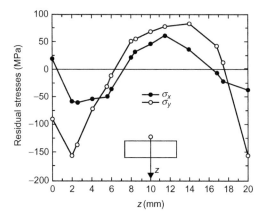

FIG. A.4 Residual stresses of TMCP (20 mm thick).

A.1.2 Residual Stress in TMCP Steel Induced by Bead Weld

Doc. No. 1.2	Residual Stress Induced by Bead Weld [1,2]			
				Material: High
Weld. Cond.	Bead Weld	Single	Medi. Thick	Strength Steel
Analysis	Thermal El-Pl analysis by FEM.			
Components	σ_x = in rolling and welding direction, σ_y = in transverse direction.			
Characteristics	On the upper surface, the computed residual stresses with and without the effect of the initial rolling stress are shown in Fig. A.5. The influence of initial rolling stress on total residual stress is schematically shown in Fig. A.6. It can be approximately expressed by the influence factor α, which is 0.0 in the welded metal and HAZ, 1.0 in the elastic deformation zone, and 0.0 to 1.0 in the plastic deformation zone, respectively.			

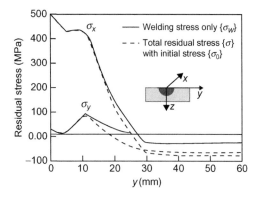

FIG. A.5 Residual stresses in TMCP induced by bead welds.

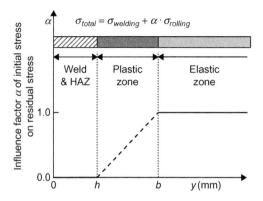

FIG. A.6 Influence factor α of initial rolling stress on residual stress.

A.1.3 Explosive Clad Steel

Doc. No. 1.3	Explosive Clad Steel [3]		
Production	Explosively Cladding	Thin	Ni/SUS304
Experiment	Measured by inherent strain method.		
Components	σ_x = in explosive direction, σ_y = transverse direction to σ_x.		
Characteristics	Very high tensile residual stress at the interface of the joint and decreases rapidly a short distance away.		

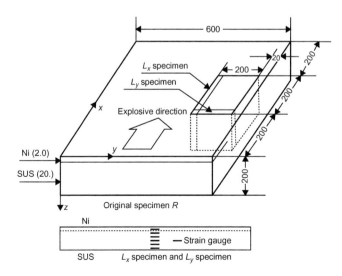

FIG. A.7 L_x and L_y specimens for measurement (Ni/SUS304 explosive clad steel).

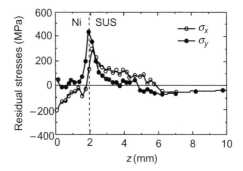

FIG. A.8 Measured residual stress at transverse cross-section (*L* method of inherent strain method).

A.1.4 Cylindrical Thick Plate by Cold Bending

Doc. No. 1.4	Cylindrical Thick Plate by Cold Bending [4]		
Manufacture	Cold Bending	Thick Plate	80 kgf/mm² HT
Experiment	Inherent strain method.		
Components	σ_θ^M = circumferential, σ_z^M = axial.		
Characteristics	Similar to residual stress distribution of a beam subjected to elastic-plastic cold bending.		

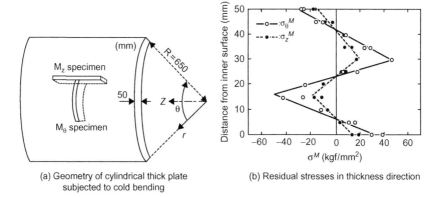

(a) Geometry of cylindrical thick plate subjected to cold bending

(b) Residual stresses in thickness direction

FIG. A.9 Residual stresses due to cold bending in shell plate.

A.2 RESIDUAL STRESSES IN WELDED JOINTS OF PLATES; IN 2-DIMENSIONAL

A.2.1 Butt-Welded Joints; Classification of Patterns of Residual Stress Distributions

Doc. No. 2.1	Classification of Patterns of Residual Stress Distributions [5,6]			
			Submerged Arc	
Welding	Butt Joint	Single	Welding	Mild Steel
Analysis	Thermal el-pl analysis by FEM.			
Component	σ_x = in welding direction.			
Characteristics	Transverse distributions of longitudinal residual stress may be classified into typical three types (A, B, C). Each pattern depends on L/B and distance from the ends. Summary is shown in figures.			

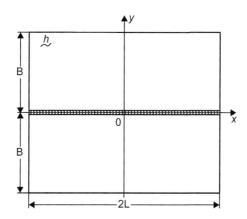

FIG. A.10   Butt-welded joint.

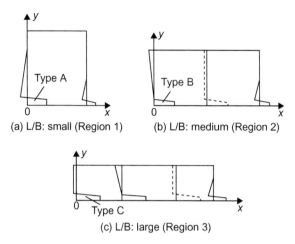

FIG. A.11 Change of stress distributions with variation of L/B (Figs. A.12 and A.13).

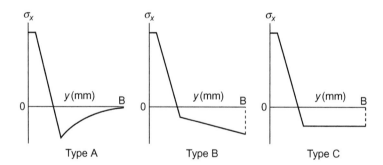

FIG. A.12 Types of stress distribution patterns.

The types of transverse distribution of longitudinal residual stress depends on L/B and the distance from the ends. A change of pattern is influenced by the distance from the ends as shown in Figs. A.11 and A.13 and summarized in Table A.1.

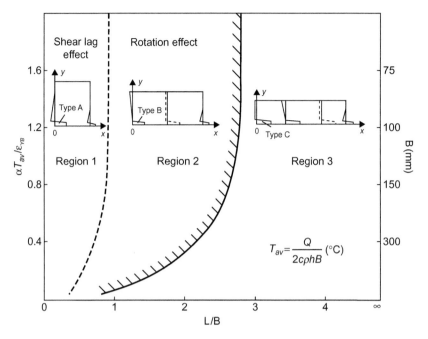

FIG. A.13 Classification of patterns of residual stress distribution.

TABLE A.1 Classification of Longitudinal Residual Stresses

Region	L/B	Transverse Distribution		
		Close to the Ends	Between Ends and Middle	Middle Portion
1	Small	Type A	Type A	Type A
2	Medium	Type A	Type C	Type B
3	Large	Type A	Type B	Type C

Type A: Effect of shear lag from the ends: large.
Type B: Effect of in-plane bending: large.
Type C: Effect of shear lag and in-plane bending is compensated or vanishes.

A.2.2 Long Butt-Welded Joint, Prediction Equation

Doc. No. 2.2	Predicting Method of Residual Stress of Long Welded Joints [7]			
Welding	Butt	Single	Submerged Arc	Mild Steel
Theoretical prediction	Longitudinal residual stress distribution in transverse cross-section at the middle length (based on theoretical analysis).			
Component	σ_x = in welding direction.			
Characteristics	Predicting equation for inherent strain distribution in transverse section. Residual stress is calculated by elastic analysis regarding inherent strain as initial strain. Equation is a function of L/B, heat input, and kind of steel.			

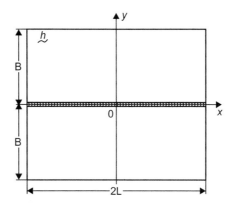

FIG. A.14 Butt-welded joint.

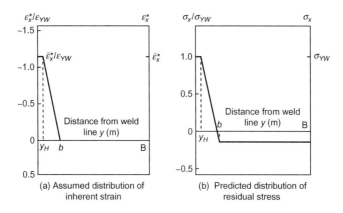

(a) Assumed distribution of inherent strain

(b) Predicted distribution of residual stress

FIG. A.15 Assumed inherent strain and predicted residual stress distributions.

The prediction for the inherent strain region is:

$$\varsigma = \bar{\varepsilon}_x^* / \varepsilon_{YW} = \left(1 + \frac{0.27\alpha T_{av}}{\varepsilon_{YB}} \right)$$

$$\xi = b/b_0 = 1 - \frac{0.27\alpha T_{av}}{\varepsilon_{YB}},$$

where

$\bar{\varepsilon}_x^*$: maximum inherent strain in the welded zone
$\varepsilon_{YW}, \varepsilon_{YB}$: yield strains of welded zone (weld metal and HAZ) and base metal
Yield stress divided by Young's modulus:
b: half width of inherent strain zone
b_0: half width of inherent strain zone in case of infinitive plate width
T_{av}: average temperature increase

$$b_0 = \frac{\alpha Q}{\sqrt{2\varepsilon\pi e c\rho h \varepsilon_{YB}}}$$

$$T_{av} = \frac{Q}{2c\rho h B}$$

A.2.3 Built-Up Members of T Shape and I Shape

Doc. No. 2.3	Built-Up Members of T Shape and I Shape [7]			
Welding	Fillet	Single	Submerged Arc	Mild Steel
Prediction and FEM analysis	1. Prediction by elastic analysis (EA) using assumed inherent strain (Doc. No. 2.2). 2. Thermal-elastic-plastic analysis (TEPA) by FEM.			
Component	$\sigma_x =$ in welding direction.			
Characteristics	Prediction of welding residual stress by elastic analysis. Verification of the predicted residual stress by TEPA of FEM.			

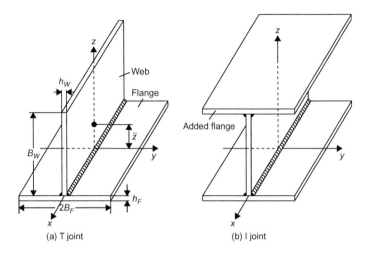

FIG. A.16 Built-up members of T shape and I shape.

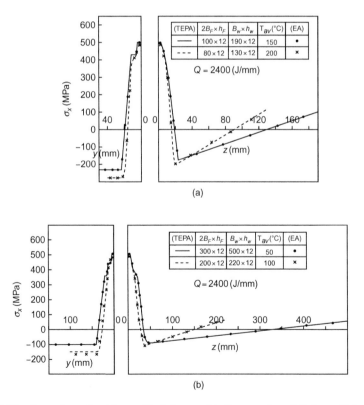

FIG. A.17 Comparison of welding residual stresses of built-up member of T shape obtained by TEPA and EA with assumed inherent strain.

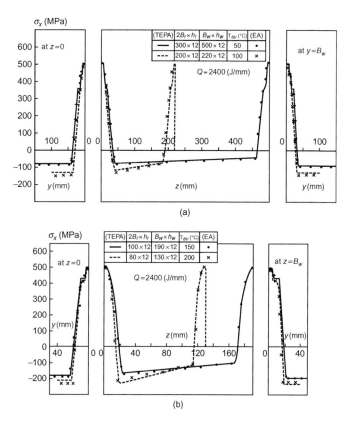

FIG. A.18 Comparison of welding residual stresses of built-up member of I shape obtained by thermal-elastic-plastic analysis (TEPA) and elastic analysis (EA) with assumed inherent strain.

A.2.4 Built-up Member of T Shape, Experiment

Doc. No. 2.4	Built-up Member of T Shape, Experiment [8]			
Welding	Fillet	Single	Submerged Arc	Mild Steel
Experiment, analysis, and prediction	1. Measurement by inherent strain method. 2. TEPA by FEM. 3. Prediction by EA using assumed inherent strain (Doc. No. 2.2).			
Component	σ_x = in welding direction.			
Characteristics	1. Establishment of measuring method by inherent strain method. 2. Comparison of welding residual stresses of built-up member of T shape obtained by TEPA and EA with assumed inherent strain.			

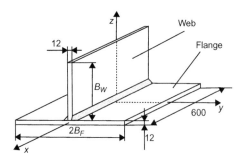

FIG. A.19 Built-up member of T shape by fillet weld (in mm).

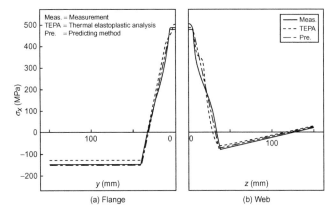

FIG. A.20 Comparison of welding residual stresses by measurement and theoretical analyses (TEPA and EA with assumed inherent strain).

A.2.5 Residual Stress and Inherent Displacement Induced by Slit Welds

(Refer to Section 7.2, "Cold Cracking of Slit Weld.")

Doc. No. 2.5	Residual Stress and Inherent Displacement Induced by Slit Welds [9]		
Welding	Slit Welds	Single	Medium Thickness SS41
Experiment and analysis	1. Experiment. 2. Elastic-plastic analysis (EPA) by FEM with inherent displacement.		
Component	σ_x = in welding direction, S_T = transverse inherent displacement.		
Characteristics	1. Measurement of residual stress and inherent displacement in slit weld cracking specimen. 2. Effectiveness of elastic-plastic analysis with inherent displacement.		

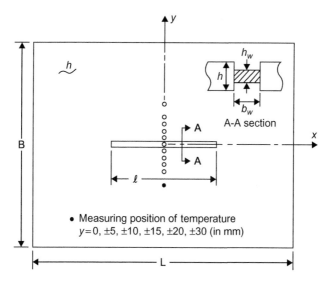

FIG. A.21 Specimen with slit welds.

TABLE A.2 Size of Specimens Furnished for Experiment [9]

No	B (mm)	L (mm)	l (mm)	h (mm)	h_w (mm)	Groove	Q (J/cm)	σ_Y	σ_u	Note
1	150	200	80				2950			
2	300	200	80				3100			
				3.2		I		46	57	SS41
3	150	400	160				3050			
4	300	400	160				3050			
5	150	200	80	25	5	Y	16500			SM41A

σ_Y: yield strength, σ_u: tensile strength of weld metal (kg/mm^2).

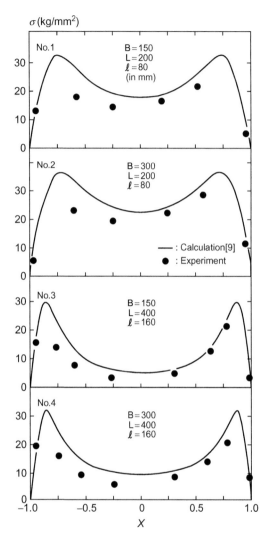

FIG. A.22   Welding residual stress distributions (experiment and analysis).

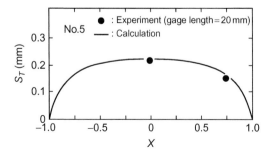

FIG. A.23   Inherent displacements S_T (experiment and calculation).

A.3 MULTIPASS BUTT WELDS OF THICK PLATES; 3-DIMENSIONAL

A.3.1 Multipass Butt Welds of Thick Plates, Classification

(Refer to Section 7.5, "Analysis of Transient and Residual Stresses of Multipass Butt Welds of Thick Plates in Relation to Cold Cracks, Under-bead Cracks, Etc.")

Doc. No. 3.1	Multipass Butt Welds, Classification [10]			
Welding	Multipass (by Single Pass)	Butt Weld with Narrow Gap	Thick	SM50
Analysis	Thermal el-pl analysis by FEM.			
Component	σ_x = transverse to weld line, σ_z = in weld line, τ_{xy} = in transverse section.			
Characteristics	1. Production mechanism of residual stresses under completely constrained and free conditions. 2. Proposal of simplified specimens for examination of weld cracks near finishing surface for both experimental and theoretical analysis).			

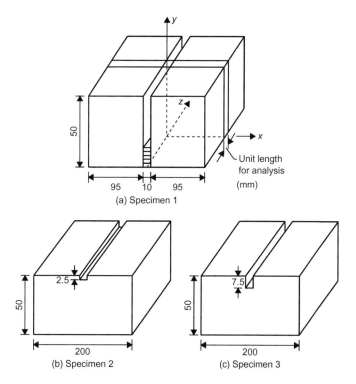

(a) Specimen 1

(b) Specimen 2   (c) Specimen 3

FIG. A.24 Specimens for analysis.

A.3.2 Multipass Butt Welds of Thick Plate, Experiment

Doc. No. 3.2	Multipass Butt Weld of Thick Plate [11]		
Welding	Multipass Butt Weld Submerged Arc Thick SS41		
Experiment	Measurement by inherent strain method.		
Component	σ_x = in welding direction, σ_y = transverse to welding direction, σ_z = in thickness direction.		
Characteristics	1. The first application of inherent strain method. 2. Measurement of three-dimensional residual stresses.		

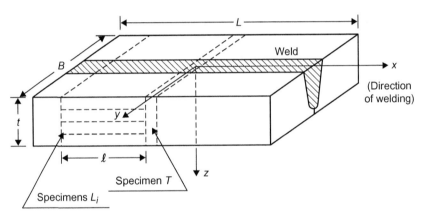

(a) Welded joint of thick plate (Specimen R)

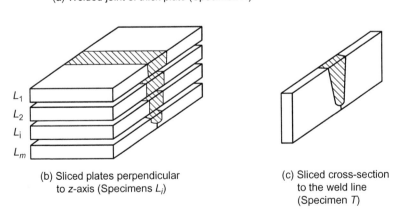

(b) Sliced plates perpendicular to z-axis (Specimens L_i)

(c) Sliced cross-section to the weld line (Specimen T)

FIG. A.25 Specimen R for measurement of three-dimensional residual stresses and specimens T and L_i ($L = 200$, $B = 200$, $t = 50$, $l = 70$ (in mm)).

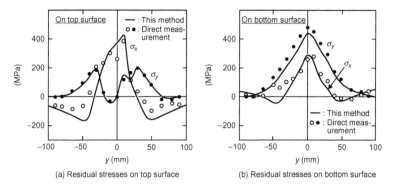

FIG. A.26   Three-dimensional residual stresses on top and bottom surfaces in the middle section.

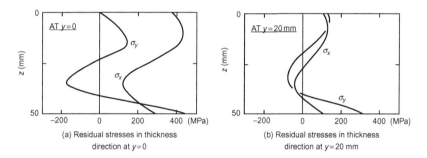

FIG. A.27   Three-dimensional residual stresses on the middle cross-section ($X = 0$).

A.4 ELECTRON BEAM WELDING, THICK PLATE

Doc. No. 4.1	Electron Beam Weld Through Thickness, Thick Plate [12]				
Welding	Butt Weld	Through Thickness	Electron Beam	Thick	HT SM50
Experiment	Inherent strain method.				
Component	σ_x = in welding direction, σ_y = in transverse direction to weld line, σ_z = in thickness, $\hat{\sigma}_x$ = most probable value, $_m\sigma$ = directly measured.				
Characteristics	1. σ_x on weld line is large. 2. There is some difference between σ_x distributions in thickness direction estimated by both measuring methods.				

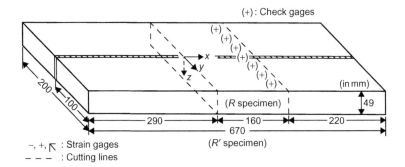

FIG. A.28 *R'* specimen jointed by butt weld through thickness (measurement was conducted on *R* specimen).

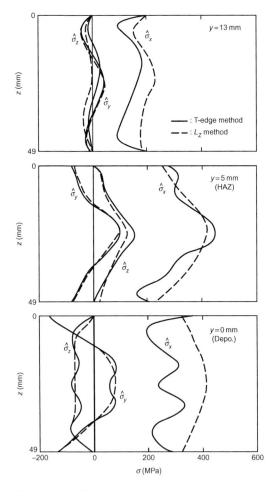

FIG. A.29 Three-dimensional welding residual stress distributions on cross-section due to electron beam welding (at $y = 0, 5, 13$ mm).

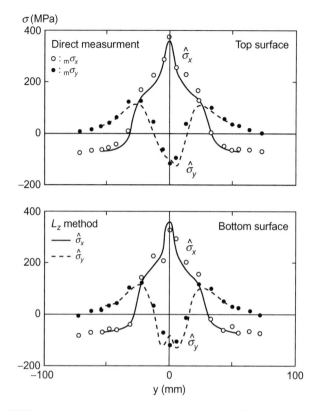

FIG. A.30 Welding residual stresses on top and bottom surfaces of R' specimen.

A.5 FIRST BEAD OF BUTT JOINT; RCC (RIGIDLY RESTRAINED CRACKING) TEST SPECIMEN

(Refer to Section 7.1, "Cold Cracking at the First Pass of a Butt-Welded Joint Under Mechanical Restraint.")

Doc. No. 5.1	First Bead of Butt Joint; RRC Test [13,14]				
Welding	Butt	First Bead	Manual	Medium	Mild Steel
Analysis	Thermal el-pl analysis by FEM.				
Component	σ_x = transverse to weld line.				
Characteristics	1. Analyses for rigidly restrained and free conditions. 2. Maximum tensile stress is produced at the root of the weld metal and may cause root cracking.				

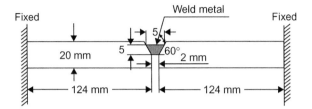

FIG. A.31   Rigidly restrained cracking specimen (RRC specimen).

A.6 MULTIPASS-WELDED CORNER JOINT

(Refer to Section 7.4, "Multipass-Welded Corner Joints and Weld Cracking.")

Doc. No. 6.1	Multipass-Welded Corner Joint of Several Groove Shapes [15]				
Welding	Corner Joint Multipass SMAW Medium HT SM50				
Analysis and experiment	1. Thermal el-pl analysis. 2. Measurement by strain gage.				
Component	σ_x = transverse, K_B = intensity of bending restraint.				
Characteristics	1. Residual stresses under various restraint conditions by CJC test apparatus. 2. Suggested measures for prevention of lamellar tearing.				

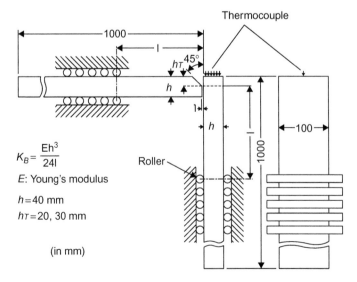

$$K_B = \frac{Eh^3}{24l}$$

E: Young's modulus

h = 40 mm

h_T = 20, 30 mm

(in mm)

FIG. A.32   Apparatus for corner joint cracking test (CJC test).

A.7 FILLET WELDS: 3-DIMENSIONAL

A.7.1 Single Fillet Welds

Doc. No. 7.1	Single Fillet Weld [16]			
Welding	Single Fillet (Leg Length 7~8 mm)	SMAW	Medium	SUS316L
Experiment	Inherent strain method.			
Component	3-D: σ_x = longitudinal (weld line), σ_y = transverse, σ_z = transverse.			
Characteristics	1. Measurement of 3-D residual stresses at fillet weld.			
	2. Estimation of functional expression of inherent strain distribution.			

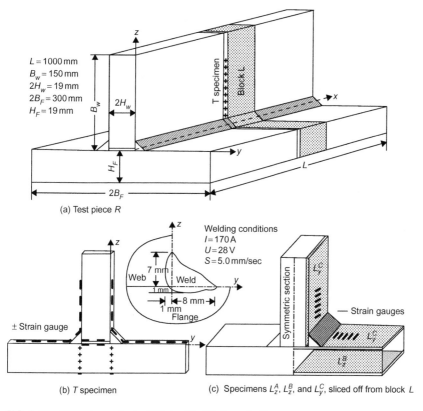

$L = 1000$ mm
$B_w = 150$ mm
$2H_w = 19$ mm
$2B_F = 300$ mm
$H_F = 19$ mm

(a) Test piece R

Welding conditions
$I = 170$ A
$U = 28$ V
$S = 5.0$ mm/sec

(b) T specimen

(c) Specimens L_z^A, L_z^B, and L_y^C, sliced off from block L

FIG. A.33 Measuring method of fillet welds (TL_yL_z method).

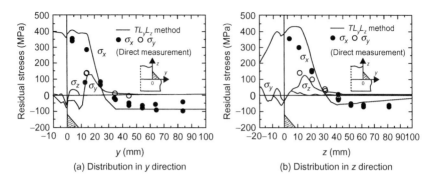

FIG. A.34 Measurement by TL_yL_z method of inherent strain method and directly observed value of residual stress at fillet welds.

A.7.2 Fillet Welds at the Joint of Web and Flange

(Refer to Section 7.3, "Analysis of Welding Residual Stress of Fillet Welds for Prevention of Fatigue Cracks")

Doc. No. 7.2	Fillet Welds at the Joint of Web and Flange [17,18]
Welding	Joint of Web and Flange Single Fillet Manual SM40
Experiment and analysis	1. Direct measurement by strain gages. 2. Thermal el-pl analysis by FEM.
Component	σ_x = in welding direction, σ_y = transverse to weld line.
Characteristics	1. Generally, σ_x is larger. 2. σ_x is largest at toe and root.

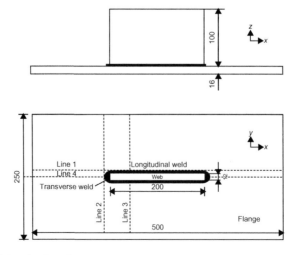

FIG. A.35 Joint of web to flange.

A.8 REPAIR WELD OF THICK PLATE

Doc. No. 8.1	Repair Weld of Thick Plate [19]			
Welding	**One Pass Repair Weld**	Submerged Arc	Thick	Mild Steel
Analysis	Thermal el-pl analysis by FEM.			
Component	σ_x = in welding direction, σ_y = transverse to σ_x, σ_z = in thickness direction.			
Characteristics	1. To remove the assumed weld defect and put one pass weld. 2. Large tensile residual stress produced at the bottom of weld metal.			

$L = 600$
$B = 300$
$h = 90$ (in mm)
$\ell = 100$
$b = 8$
$d = 5$

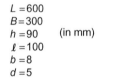

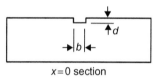

$x=0$ section

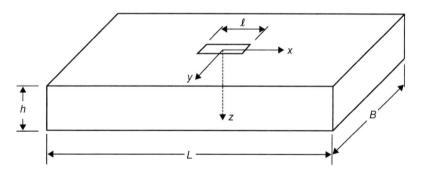

FIG. A.36 Repair welding model for analysis.

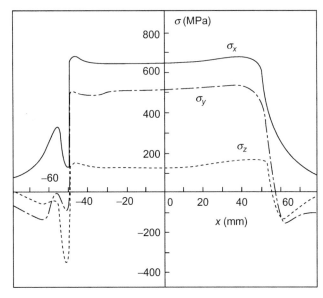

(a) 3-D residual stress

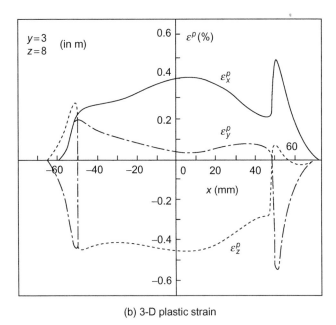

(b) 3-D plastic strain

FIG. A.37 Distributions of three-dimensional residual stresses and plastic strains along weld line.

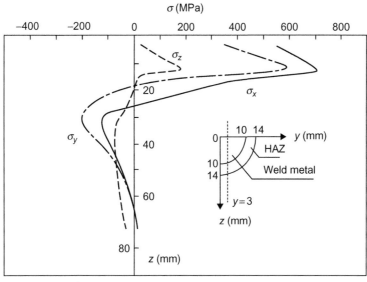

(a) 3-D residual stress

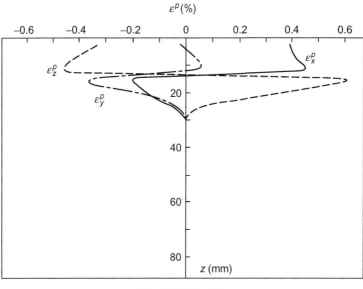

(b) 3-D plastic strain

FIG. A.38 Distributions of three-dimensional residual stresses and plastic strains in z direction (at middle section).

A.9 CIRCUMFERENTIAL WELDED JOINT OF PIPES

A.9.1 Circumferential Welded Joint of Pipes—Heat-Sink Welding

(Refer to Section 7.6, "Improvement of Residual Stresses of Circumferential Joint of a Pipe by Heat-sink Welding")

Doc. No.9.1	Circumferential Welded Joint of Pipes Heat-sink Welding [20]			
Welding	Multi-pass	1. Conventional welding 2. Heat-sink welding	Thin, Medium, Thick	SUS 304
Analysis	Thermal elastic-plastic analysis by FEM			
Component	σ_z = axial σ_θ = circumferential			
Characteristics	1. Heat-sink welding is effective to produce compressive residual stress on inner surface of the pipe. 2. The method can be more effective by providing lower heat input for final few passes of thin pipe and for first few passes of thick pipe.			

TABLE A.3 Welding Condition

Pass No. \ Size of pipe	2B	4B Q - 14	4B Q - 23	4B Q - 45	24B
1	8.2		13.3		7.2
2	8.2		7.3		21.7
3	8.2		7.3		13.1
4	8.2	15.3	18.0	25.0	13.1
5	8.2	14.2	22.0	45.0	13.1
6	8.2	14.0	23.0		13.1
7 ~ 16					19.8

(KJ/cm)

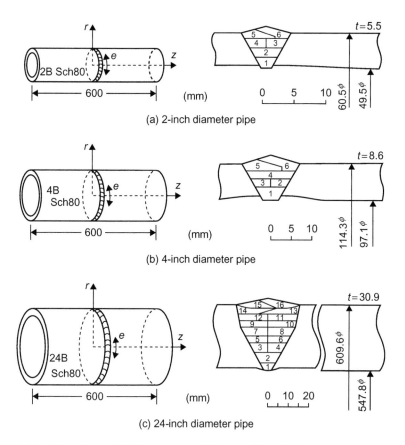

(a) 2-inch diameter pipe

(b) 4-inch diameter pipe

(c) 24-inch diameter pipe

FIG. A.39 Dimensions and built-up sequences of pipes used in analysis.

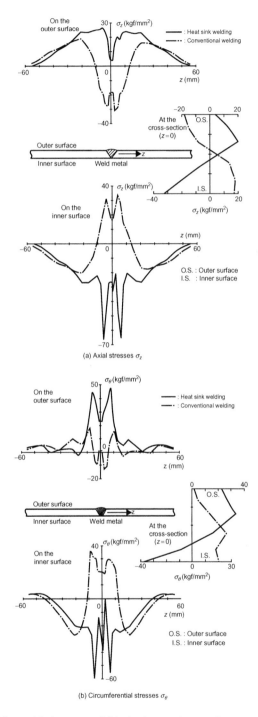

FIG. A.40 Welding residual stresses of 2-inch pipes on inner and outer surfaces at the cross section ($z = 0$).

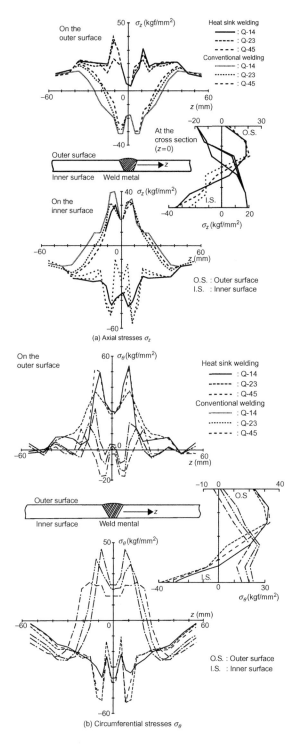

FIG. A.41 Welding residual stresses of 4-inch pipes on inner and outer surfaces at the cross section ($z = 0$).

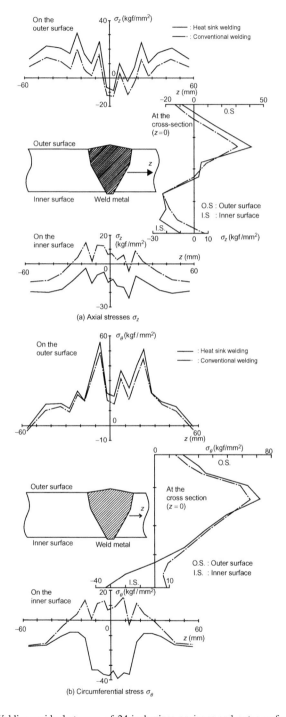

FIG. A.42 Welding residual stresses of 24-inch pipes on inner and outer surfaces at the cross section ($z = 0$).

A.9.2 Penetrating Pipe Joints in Nuclear Reactor

Doc. No.9.2	Penetrating Pipe Joint [21, 22, 23]			
Welding	Multi-pass (21 passes)	Circumferential	Thin to thick	SUS316, SM490, ENiCrFe-1
Experiment & analysis	1. Direct measurement by strain gages 2. Measurement by inherent strain method			
Component	σ_θ=circumferential σ_r=radial			
Characteristics	1. Tensile stress around joint 2. Tensile stress in vessel			

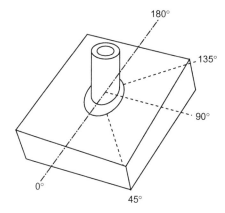

FIG. A.43 Penetrating welded joint of pipe and plate.

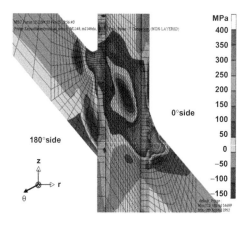

FIG. A.44 Residual stress (σ_θ) in cross section at 0°–180°.

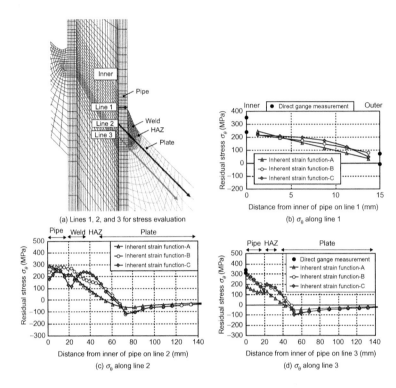

(a) Lines 1, 2, and 3 for stress evaluation

(b) σ_θ along line 1

(c) σ_θ along line 2

(d) σ_θ along line 3

FIG. A.45   Circumferential residual stress (σ_θ) in cross section at 0°–180° along Lines 1, 2 and 3.

REFERENCES

[1] Ma N, Murakawa H, Ueda Y. Effects of initial stresses on welding deformation and residual stresses. Proc. 6th Int. Weld. Symp. on the Roles of Welding Science and Technology in the 21st Century, Vol. 2. Nagoya: Japan Welding Society; 1996, pp. 557–62.

[2] Ma N, Murakawa H, Ueda Y. Effect of rolling stresses on welding residual stresses. Proc. of Annual Meeting, Vol. 57. JWS, 1995, pp. 186–87 (in Japanese).

[3] Ueda Y, Murakawa H, Ma N. Measuring method for residual stress in explosively clad plates and a method of residual stress reduction. ASME, Jl Engi Mater Tech 1996;118(10):576–82.

[4] Ueda Y, Fukuda K, Nishimura I, Iiyama H, Chiba N, Fukuda M. Three-dimensional cold bending and welding residual stresses in penstock of 80kgf/mm^2 class high strength steel plate. Trans JWRI 1983;12(2):117–26.

[5] Ueda Y, Kim Y, Yuan M. A predicting method of welding residual stress using source of residual stress (Report I), characteristics of inherent strain (Source of Residual Stress). Trans JWRI 1989;18(1):135–41.

[6] Yuan, M. Research on estimating method of welding residual stress in plated structures based on the characteristic of inherent strain. Ph.D. Dissertation, Osaka Univ., Japan; 1990.

[7] Ueda Y, Yuan M. A predicting method of welding residual stress using source of residual stress (Report II), determination of standard inherent strain. Trans JWRI 1989;18(1):143–50.

[8] Ueda Y, Yuan M, Mochizuki M, Umezawa S, Enomoto K. A predicting method of welding residual stress using source of residual stress (Report IV), experimental verification for predicting method of welding residual stresses in T-joints using inherent strains. Trans JWRI 1993;22(1):179–86.

[9] Ueda Y, Fukuda K, Kim Y, Koki R. Characteristics of restraint stress-strain of slit weld in a finite rectangular plate and the significance of restraint intensities as a dynamical measure. Trans JWRI 1982;11(2):105–13.

[10] Ueda Y, Nakacho K. Simplifying methods for analysis of transient and residual stresses and deformations due to multipass welding. Trans JWRI 1982;11(1):95–103.

[11] Ueda Y, Fukuda K, Tanigawa M. New measuring method of 3-dimensional residual stresses based on theory of inherent strain. Trans JWRI 1979;8(2):249–56.

[12] Ueda Y, Kim Y, Umekuni A. Measuring theory of three-dimensional residual stresses using a thinly sliced plate perpendicular to welded line. Trans JWRI 1985;14(2):151–57.

[13] Ueda Y, Yamakawa T. Analysis of thermal elastic-plastic stress and strain during welding by finite element method. Trans JWS 1971;2(2):90–100.

[14] Ueda Y, Fukuda K, Nakacho K. Analysis of welding stress by finite element method and mechanism of production of residual stresses. Jl Japan Welding Society 1976;45(1):29–35.

[15] Ueda Y, Nishimura I, Iiyama H, Chiba N, Fukuda K. Prevention of lamellar tearing in multipass welded corner joint. Trans JWS 1978;9(2):128–33.

[16] Ueda Y, Ma N, Wang Y, Koki R. Measurement of residual stresses in single-pass and multipass fillet welds using inherent strains-estimating and measuring methods of residual stresses using inherent strain distribution described as functions (Report 5). Q J JWS 1995;13(3):470–78 (in Japanese).

[17] Wu A, Ma N, Murakawa H, Ueda Y. Effect of welding procedures on residual stresses of T-Joints. Trans JWRI 1996;25(1):81–89.

[18] Ma N, Ueda Y, Murakawa H, Maeda H. FEM analysis of 3-D welding residual stresses and angular distortion in T-type fillet welds. Trans JWRI 1995;24(2):115–22.

[19] Ueda Y, Kim Y, Garatani K, Yamakita T, Bang H. Mechanical characteristics of repair welds in thick plate (Report I), distributions of three-dimensional welding residual stresses and plastic strains and their production mechanisms. Trans JWRI 1987;15(2):187–96.

[20] Ueda Y, Nakacho K, Shimizu T, Ohkubo K. Residual stresses and their mechanisms of production at circumferential weld by heat-sink welding. J JWS 1983;52(2):90–97 (in Japanese).

[21] Nayama M, Ohta T, Iwamoto Y, Ogawa N, Ishii K, Ushio M, Nakacho K. Development of innovative measuring method for three-dimensional welding residual stresses and detail study on crack propagation in welded structures, 2004–2005. http://www.iae.or.jp/KOUBO/innovation/theme/pdf/h15-1-16.pdf (in Japanese).

[22] Nakacho K, Ma N, Suwa T, Ohta T, Ogawa N. FEM program of 3-D inherent strain function for measuring welding residual stresses. Proc. of Annual Meeting. Japan: JWS; (2006); pp. 220–21.

[23] Nakacho K, Ohta T, Ogawa N, Ma N, Hamaguchi H, Satou M, Nayama M. Measurement of welding residual stresses of reactor vessel by inherent strain method—measurement of residual stresses of pipe-plate penetration joint. Trans JWS 2007;25(4):581–89.

Contents of Programs and Data on the Companion Website

The companion website is located at booksite.elsevier.com/9780123948045

- **Appendices**
  【Appendix-B】 : Published material properties for thermal elastic-plastic
  FEM analysis
  【Appendix-C】 : Theory of three-dimensional thermal elastic-plastic creep
  analysis
- **Folders and distributed programs, documents**
  【PROG】
  heat2d.exe
  tepc2d.exe
  inhs2d.exe
  awsd.exe
  【Manual】
  install_manual.pdf
  head2d_manual.pdf
  tepc2d_manual.pdf,
  inhs2d_mamual.pdf
  post_manual.pdf
  【License】
  JWRI.FILE
- **Folders list of sample data**
  【DATA1】 : Thermal stress cycle in one-dimensional constraint bar with
  $T_{max} = 50°C$
  【DATA2】 : Thermal stress cycle in one-dimensional constraint bar with
  $T_{max} = 150°C$
  【DATA3】 : Thermal stress cycle in one-dimensional constraint bar with
  $T_{max} = 300°C$
  【DATA4】 : Thermal stress cycle in one-dimensional constraint bar with
  $T_{max} = 600°C$
  【DATA5】 : Thermal stress cycle in one-dimensional constraint bar with
  $T_{max} = 600°C$ and 10h creep behavior
  【DATA6】 : Heat conduction sample using moving welding heat source
  【DATA7】 : Heat conduction sample using fixed welding heat source
  【DATA8】 : Slit welding heat conduction sample using fixed welding heat
  source

【DATA9】 : Slit welding thermal stress sample using thermal elastic-plastic FEM

【DATA10】 : Slit welding residual stress sample using inherent strain FEM (reproduction of residual stresses by residual plastic strains)

【DATA11】 : Butt welding heat conduction sample using moving welding heat source

【DATA12】 : Butt welding thermal stress sample using thermal elastic-plastic FEM

【DATA13】 : Butt welding residual stress sample using inherent strain method (Reproduction of residual stresses and deformation by plastic strains)

【DATA14】 : Butt welding residual stress sample using inherent strain method (Numerical experiment for residual stress measurement)

【DATA15】 : Butt welding residual stress sample using inherent strain method (Estimation of residual stress using predicted inherent strain)

【DATA16】 : Thermal cutting heat conduction sample using moving heat source

【DATA17】 : Thermal cutting deformation sample using thermal elastic-plastic FEM

【DATA18】 : Thermal cutting residual deformation using inherent strain method (Reproduction of deformation by residual plastic strains)

【DATA19】 : Thick plate bead welding heat conduction

【DATA20】 : Thick plate bead welding induced thermal stresses and deformation using thermal elastic-plastic FEM

【DATA21】 : Thick plate bead welding induced residual stresses and deformation (Reproduction of residual stresses and deformation by plastic strains)

Index

Page numbers in *italics* indicate figures and tables

A

Average cooling rate, on weld line, 178
Average rate of temperature change, 139
awsd.exe, 132, 144
Axial residual stresses, 212

B

Bar C
  with elastic restraint, *22*, 28
  fixed at both ends, 13, *14*
    high temperature heating process, 18–21
    low temperature heating process, 15–16
    medium temperature heating process, 16–18
  free bar, 12–13, *12*
  mechanical conditions, 11, *11*
  movable rigid body
    high temperature heating process, 26–27
    low temperature heating process, 22–24
    medium temperature heating process, 24–26
Base metal, 57
Basic theoretical solutions, 115–122
Book's website, samples data on, *153–154*
  DATA6 and DATA7, 155
  DATA1-DATA5, 152, *153–154*
  DATA8-DATA10, 157
  DATA11-DATA15, 163
  DATA16-DATA18, 163–164
  DATA19-DATA21, 164, 166
Boundary conditions, 110–112
  input data for, 190–191, 195–196, 202
  mesh division and, *194*, *201*
Butt welding
  moving heat source for, *175*
  of pipes, *82*
  of thin plate, *80*
  transient temperature distribution during, *133*
Butt-welding joints, *170*, 170, 178, 253–255
  dimensions of, *189*
  long, prediction equation, 256–257

temperature distribution on thick plate, 119–120
thin plate
  inherent strain distribution on, 120–122
  temperature distribution on, 116–119

C

CCT, *see* Continuous cooling transformation
check.txt, 139, 141, 143
CJC, *see* Corner joint weld cracking
Computational Welding Mechanics, 8
Concentrated heat source, 115–116
Connected pipe model, *77*
Continuous cooling transformation (CCT), 68
Control parameters
  of heat conduction computation, 175–176
  thermal elastic-plastic creep computation, 183–185
Conventional methods, of elastic strain measurement, 40, *40*
conv.txt, 141
Cooling rate contour, 148, *148*
Cooling speed, 73
Corner joint weld cracking (CJC), 221
Creep deformation, 92–93
Creep strain increment, 92–93
Cylindrical thick plate, by cold bending, 252

D

Deformation
  elastic, 89
  FEM program, execution steps, 199
  prediction of, 235–236
  welding shrinkage, 187
    distribution of, 192–193
Diffusion, 56–59
Displacement–strain relation, 38–39, 105
disp.txt, 141, 143
Dissipation, 56–59

E

Effective inherent strains, 52, 197
  and noneffective, 41–42
  and residual stress measurement
    accuracy of, 45–46
    determination, 42–45
  zone, *196*
ehist.txt, 141
Elastic deformation, 89
Elastic distribution, 214
Elastic response equation, 50–51
Elastic response matrix, derivation of,
    46–48
Elastic strains, 40, 44, *196*
  distribution of, *121*
  increments, 89
  measuring methods for, *40*, 40, 198
Elastic-plastic distribution, 215
Electric current analysis, 103
Electric power, 2, *4*
Electron beam welding, of thick plate,
    265–267
Equations, for types of problems, *102*
Equilibrium condition, 39–40
Equilibrium equation, 86–87, 95
  of bar, 104–105
  of two bars, 105–108
estrain.txt, 141, 143
Explosive clad steel, 251–252

F

FEM, *see* Finite element method
Fillet welding, *81*
  division of, *219*
  residual stress welding in, *218*
  at web and flange joint, 270
Finite element method (FEM), 1
  as powerful tool for problems, 99–101
  program, 131
    execution steps, 185
    features in graphical representation, *132*
    inherent strain, simulation steps using,
      188–193
    temperature distribution on
      butt-welded joint thick plate, 119–120
      butt-welded joint thin plate, 116–119
      concentrated heat source, 115–116
Fixed welding heat source, *136*, 136
Fusion welding, 1

G

Groove shapes, types of, 2, *4*

H

HAZ, *see* Heat-affected zone
Heat conduction, 57, 63
  analysis, 124
    checklist for, 125
    troubleshooting in, 126–128
  computation
    control parameters of, 175–176
    output files after, 139, *139*
    program features in, *132*
    time step for heat conduction, 138
  simple method for solving problem, 74–79
Heat supply, 56–59
Heat transfer, 58, 137
Heat-affected zone (HAZ), 2, 68
heat2d.exe, 131, 135, 176
Heating time, *see* Welding time
Heat-sink welding, 231–235
Hole drilling method, 35

I

I shape, built-up members of, *257*, *258*
Idealized model, for residual stress, 9–10
Incremental method, for nonlinear problems,
    108–109
Independent inherent strains, 197
Inherent displacement, 18
  high temperature heating process, 33–34
  induced by slit welds, 260–262
  low temperature heating process, 30–31
  medium temperature heating process, 31–32
Inherent strains, 197
  analysis, checklist, 126
  based FEM program, 131, *134*, 142–143
  components, 191, 197, 202
  concept of, 8
  distribution function, *199*, 202–203
    parameters of, 203–204
  effective
    and noneffective, 41–42
    residual stress measurement, 42–45
  experiments and, measured strains in, 40–41
  high temperature heating process, 33–34
  idealized model, 9–10
  input data of, 191–192, 205–206
  input file preparation, 200–206
  inverse analysis for, 28–29
  low temperature heating process, 30–31
  medium temperature heating process, 31–32
  method, 35
  parameters, calculation of, 204–205
  prediction formula of, 200–206

reproduction of residual stress by, 27–28
and resulting stresses, 36–40
temperature distribution, 5, 8–9
and thermal elastic-plastic method,
   192–193, 199–200, 206–207
unknown number of, 197
zone, 196–197
inhs2d.exe, 131, *134*, 142–143
inhs2d.post, 143
Input data, saving, 198, 206
Instantaneous heat source, 115, *116*
Inverse analysis, for inherent strain, 28–29
istrain.txt, 143

L

Longitudinal residual stress, 186–187, *212*, *255*

M

Manual arc metal (MMA), 221
Material properties
   aluminum and mild steel, *64*
   changes with temperature, 64–69
   differences in, 63–64
   used in heat conduction analysis, *172*,
      172–173
   used in inherent strain FEM program, 190,
      195, 201
   used in thermal elastic-plastic simulation,
      180–182, *181*
Mechanical boundary conditions, 159, 163
Mechanical constraint, 8–9
Mechanical melting point, 55, 70
Mesh division, 109–110
   and boundary conditions, *194*, *201*
   input data for, 171–172, 189–190,
      194–195, 201
   thermal elastic-plastic FEM program, *189*, 189
Metallurgical melting temperature, 70
MMA, *see* Manual arc metal
Moving heat source model, 135, *136*
Multipass butt welds, of thick plates, 263–265
Multipass fillet welds, *220*
Multipass-welded corner joints, 221–227, 268

N

Newton's law, 58
nhist.txt, 141
Nodal displacements, 104
Nodal forces, 104
Nodal temperature, 99
Node, 99
   temperature file, 185

Noneffective inherent strains, 41–42
Nonlinear problems, incremental method for,
   108–109

O

Observation equation, 40

P

Plastic deformation, 89–92
Plastic distribution, 215
Plastic strain, 10, 20, 26, 34
   contour, 149, *150*
   increments, 89–92
Poisson's ratio, *68*
Post program, 185–188
Post-processing program awsd.exe
   cooling rate contour, *148*, 148
   inherent strain contour, 150, *151*
   stress contour, 149, *149*
   stress distribution graph, 150, *151*
   temperature contour, 147, *148*
   use, 144
Predicted welding distortion, *243*
Prediction, 6
   welding residual stresses, 52
Principle of minimum potential energy, 104
Principle of virtual work, 104
Program installation, 133
pstrain.txt, 141

R

Rate of temperature change, 138, *138*
Rational simulation, checklist for, 124–126
Residual plastic strain, distribution of, 187–188
Residual stress, 35, 36, 194
   computation, steps to execute inhs2d.exe
      for, 206
   distributions, 121, 186–187, 193, *199*, 212,
      230, *233*, *234*, 260–262
   high temperature heating process, 33–34
   idealized model, 9–10
   low temperature heating process, 30–31
   medium temperature heating process,
      31–32
   patterns classification of, 253–255
   temperature, 5, 8–9
   thermal elastic-plastic analysis, 10–12
   in TMCP, 248–251
   typical patterns of, 6, *7*
   in typical welded joints, 272–277
effective inherent strains, determination of,
   42–45

Residual stress (*Cont.*)
    FEM program, steps to execute, 199
    by inherent strain method, computation
        steps, 200–207
    measurement, numerical experiment for,
        193–200
    single pass and multipass welds, 219–221
    thermal elastic-plastic analysis, 221–223
    three-dimensional analysis, 216–219
    in two and three-dimensional models, 48–52
    welding deformation and, 122–123
Resistance spot welding process, simulation of,
    236–239
Rigid body movement, constraint condition for,
    191, 196
Rigidly restrained cracking (RRC), 210
    test, 267–268
RRC, *see* Rigidly restrained cracking

S
Simple heat flow model, 59–63
    temperature distribution, 60–61
Simple post processing program, 132
Simple simulation model, residual stress, 9, *9*
Simulation
    rational, checklist for, 124–126
    steps of, 123, *123*
Single fillet welds, three-dimensional residual
    stresses of, 269–270
Slit weld
    cold cracking of, 213–216
    mechanical behavior of, 214
Spot welding simulation, flowchart of, *237*
Stefan-Boltzman's law, 58
Strain-displacement relation, 86, 94–95
Strain-hardening coefficient, 91
Strains
    hardening curve, *91*
    increment, decomposition of, 88
    setting measured, input data for, 198
Stress
    analysis, 108, *109*, 112
        checklist, 125
        troubleshooting in, 126
    component, distribution of, *185*
    contour, *149*, 149
    distribution graph, 150, *151*
    measurement, inherent strain program for, 132
    relaxation method, 35, 40
Stress-strain relation, 39, 86, 95–96, 105
stress.txt, 141, 143
Symmetric boundary conditions, 190, 195
Symmetric displacement boundary, 183

T
T shape, built-up members of, *257*, *258*,
    259–260
Temperature contour, 147, *148*
Temperature distribution, 60–61
    during welding, *176*, 176–177
Temperature-stress curve, *113*, 114
TEMP.FILE, 139
tepc2d.exe, 131, *134*, 140–141, *140*
tepc2d.post, 141
Thermal conditions
    DATA1-DATA5, 152
    DATA8-DATA10, 159
    DATA16-DATA18, 163
    DATA19-DATA21, 165
    input data for, 174, 183
Thermal conductivity, *66*
Thermal cycles, 6, 30, 112
Thermal deformation, 88–89
Thermal elastic behavior, bar fixed at both ends,
    112–113
Thermal elastic-plastic
    analysis, 10–12, 221–223
        checklist, results of, 125–126
        troubleshooting for, 128
    behavior of bar fixed, both ends, 113–114
    creep behavior of bar fixed, both ends,
        114–115
    creep FEM program
        input file preparation, 179–185
        purpose and simulation conditions,
            178–179
    creep program tepc2d.exe, 131, *134*, 140,
        140–141
    inherent strain method and, 192–193,
        199–200, 206–207
Thermal expansion ratio, *66*
Thermal reflection, 177
Thermal strain
    distribution of, *121*
    increments, 88–89
Thick plate, 119–120
    repair weld of, 271–273
Thin plate, 116–119, 120–122
Three-bar model, 8, *11*, 29, 84–87, *85*
    thermal visco-elasto-plastic problem in,
        94–96
Three-cylinder model, *60*
Three-dimensional residual stresses
    measurement of, thick plates, 49–52
    of single fillet welds, 269–270
Three-pipe model, *74*
tnmax.txt, 139

Transient temperature distribution, 116,
118–119
during butt welding, *133*
by instantaneous heat source, *116*
Transient thermal field, characteristics of, *117*
Transient thermal stresses, distribution of,
185–186
Transverse residual stress, 186
Transverse section, temperature distributions
on, 177–178
Transverse stress component, 185
distribution of, *185*
Transverse welding residual stress, *226, 227*
Troubleshooting
heat-conduction analysis, 126–128
and stress analysis, 126
thermal elastic-plastic analysis, 128–129
Two-dimensional
analysis, 218
residual stresses, butt-welded joints of plate,
48–49

U

Uni-axial stress models, *37*

V

Variational principle, 103–104

W

Weld cracking, 221–227
Weld metal, 57
shrinkage of, 210
Welded joints, types of, 2, *3*
Welded structures, manufacturing processes and
classifications of, 1, 2
Welding
arc, schematic illustration of, 2, *4*
characteristic length for, *72*
conditions, *222*
butt welding, 170, 179
cracks, initiation of, 223–227

deformation
basic types of, *6*, 6
flow of analysis for, 122–123
and residual stress, 122–123
distortion, 239–244
prediction of, 239–244
dummy elements, 182–183
groups of, 4, *6*
heat conduction FEM program, 132, 135
input file preparation, 171–176, 176
purpose and simulation conditions,
170–171
viewing results using post program,
176–178
heat flow and temperature, 55–79
heat, input data for, 174–175
line, temperature distributions on, *177*,
177–178
mechanical problems, concepts of, 79–97
mechanics, physical values, 127, *127*
problems and countermeasures, 6, *7*
shrinkage deformation, 187
distribution of, 192–193
simulation, 99
stress-strain relation in, 87–94
thermal cycle, temperature distribution in,
177, *177*
thermal elastic plastic creep program, 132
time, 166
Welding residual stress distributions, *218*,
223–227
butt joint
measurement of, *194*
prediction of, *200*
characteristics of, 230–231
distribution of, *186*, 186
prediction of, 52

Y

Yield strain, 204
Yield stress, *67*
Yield temperature, 70–71

About the Authors

Yukio Ueda

Date of Birth: April 12, 1932, Osaka, Japan.

1955: B. Eng. Osaka University, Japan, 1962: PhD Lehigh University, USA, 1968; Dr. Eng. Osaka University, Japan, 2002: Honorary Dr. Norwegian University of Science and Technology, Norway.

1996 to date: Professor Emeritus, Osaka University, Japan, 1975–1996: Prof., Osaka University, 1992–1996: Director of Welding Research Institute, Osaka University, 1996–2003: Prof. Kinki University, Japan, 1977–1978: Visiting Professor, University of Michigan, USA, 1979–: Adjunct Prof. Xi'an Jiao Tong University, 1987– Adjunct Prof. Shanghai Jiao Tong University, China. 2000–2002: President, Int. Society of Offshore and Polar Engineers (ISOPE), 1994–1997: Standing Comm. of ISSC (Int. Ship and Offshore Structures Cong.), 1988–1997: Co-Editor, Marine Structures.

Hidekazu Murakawa

Date of Birth: March 15, 1951, Takamatsu, Japan.

1973: B. Eng. and 1975: M. Eng. Osaka University, Japan, 1978: PhD Georgia Institute of Tech., USA.

1978–1980: Research Associate, Georgia Tech., 1980–1983: Hitachi Research Institute, Hitachi Ltd., Japan, 1983–2001: Research Associate and Assoc. Prof., Welding Research Institute, Osaka University, 2001– Professor, Joining and Welding Research Institute, Osaka University, 2007–: Leader, Global Cooperative Research Center for Computational Welding Science, Osaka University, 2006–: Adjunct Prof. Xi'an Jiao Tong University, China.

Ninshu Ma
Date of birth: February 14, 1961, Xi'an, China.

1982: B. Eng., and 1985: M. Eng., Xi'an Jiao Tong University, China, 1993: PhD, Osaka University, Japan.

1985–1989: Assistant Professor, Xi'an Jiao Tong University, China. 1993–1996: Research Associate, Osaka University, Japan. 1996 to date: Professional, Japan Research Institute (JSOL). 2007 to date: Guest Associate Professor, Global Collaborative Research Center for Computational Welding Science, Joining and Welding Research Institute, Osaka University, Japan.

Printed and bound by CPI Group (UK) Ltd, Croydon, CR0 4YY

08/05/2025

01864890-0001